" It's amazing how many toys are based upon physics and chemistry principles. "

"Learning science concepts with toys is an exciting adventure for children. Their natural interest and curiosity in science combined with their desire to 'play' with toys provides great motivation to learn."

Jeannie Tuschl—Tulip Grove School, Nashville, Tennessee

"I really learned that there are many toys that can be used to teach science. I hope to expand the use of the TOYS concept in my classroom."

Elizabeth Henline—Mount Orab Middle School, Mount Orab, Ohio

"The toys, experiments, and activities are classroom-friendly to students of all ages."

Mary Hurst—McKinley Elementary School, Middletown, Ohio

"TOYS is a great program, with lots and lots of new and exciting ideas to use in the classroom."

JoAnne Lewis—Stanberry Elementary, King City, Missouri

"I would highly recommend TOYS for all science teachers."

Sarah Birdwell—Butterfield Junior High, Van Buren, Arkansas

"With TOYS, science really becomes part of everyday experiences and materials."

Mary White—Monmouth High School, Monmouth, Illinois

"Teaching Science with TOYS is a wonderful way to motivate children. It's a super program!"

Cindy Waltershausen—Western Illinois University, Monmouth, Illinois

"I received so many new ideas to try out in my classroom that my students will be enjoying learning science without even realizing it!"

Regina Bonamico—Chauncy Rose Middle School, Terre Haute, Indiana

"TOYS activities will enable me to help develop a love for science, genuine inquiry, and higher-level thinking skills with my students. TOYS provides a wealth of ideas to introduce hands-on learning."

Rita Glavan—St. Pius X, Pickerington, Ohio

Investigating Solids, Liquids, and Gases with TOYS

Other Teaching Science with TOYS Books by Terrific Science Press

Exploring Energy with TOYS

Exploring Matter with TOYS

Investigating Solids, Liquids, and Gases with TOYS

Teaching Chemistry with TOYS

Teaching Physics with TOYS

Investigating Solids, Liquids, and Gases with TOYS

States of Matter and Changes of State

Jerry Sarquis

Lynn Hogue

Mickey Sarquis

Linda Woodward

Terrific Science Press
Miami University Middletown
Middletown, Ohio

Terrific Science Press
Miami University Middletown
4200 East University Blvd.
Middletown, Ohio 45042
cce@muohio.edu
www.terrificscience.org

10   9   8   7   6   5   4   3

This monograph is intended for use by teachers, chemists, and properly supervised students. Users must follow procedures for the safe handling, use, and disposal of chemicals in accordance with local, state, federal, and institutional requirements. The cautions, warnings, and safety reminders associated with experiments and activities involving the use of chemicals and equipment contained in this publication have been compiled from sources believed to be reliable and to represent the best opinions on the subject as of the date of publication. Federal, state, local, or institutional standards, codes, and regulations should be followed and supersede information found in this monograph or its references. The user should check existing regulations as they are updated. No warranty, guarantee, or representation is made by the authors or by Terrific Science Press as to the correctness or sufficiency of any information herein. Neither the authors nor the publisher assume any responsibility or liability for the use of the information herein, nor can it be assumed that all necessary warnings and precautionary measures are contained in this publication. Other or additional information or measures may be required or desirable because of particular or exceptional conditions or circumstances or because of new or changed legislation.

ISBN 1-883822-28-9

The publisher takes no responsibility for the use of any materials or methods described in this monograph, nor for the products thereof. Permission is granted to copy the materials for classroom use.

This material is based upon work supported by the National Science Foundation under grant number TPE-9055448. This project was supported, in part, by the National Science Foundation. Any opinions, findings, and conclusions or recommendations expressed in this material are those of the authors and do not necessarily reflect the views of the National Science Foundation.

Contents

Program Staff

Chemistry

Mickey Sarquis
Associate Professor of Chemistry
Director, Center for Chemistry Education
Miami University Middletown
Middletown, Ohio

Jerry Sarquis
Professor of Chemistry
Miami University
Oxford, Ohio

John Williams
Associate Professor of Chemistry
Miami University Hamilton
Hamilton, Ohio

Lynn Hogue
Associate Director
Center for Chemistry Education
Miami University Middletown
Middletown, Ohio

Linda Woodward
Research Associate
Miami University Middletown
Middletown, Ohio

Physics

Beverley Taylor
Associate Professor of Physics
Miami University Hamilton
Hamilton, Ohio

Dwight Portman
Physics Teacher
Winton Woods High School
Cincinnati, Ohio

Jim Poth
Professor of Physics
Miami University
Oxford, Ohio

Mentors

Cheryl Vajda
Teacher
Stewart Elementary
Oxford, Ohio

Gary Lovely
Physics Teacher
Edgewood Middle School
Hamilton, Ohio

Tom Runyan
Science Teacher
Middletown High School
Middletown, Ohio

Terrific Science Press Design and Production Team

Susan Gertz
Document Production Manager
Illustration
Design/Layout

Amy Stander
Technical Coordinator
Technical Writing
Technical Editing

Lisa Taylor
Technical Writing
Technical Editing
Production

Stephen Gentle
Illustration
Photo Editing
Design/Layout
Production

Amy Hudepohl
Technical Writing
Technical Editing
Production

Jennifer Stencil
Design/Layout
Production

Stephanie Beaver
Production

Thomas Nackid
Illustration

Andrea Nolan
Laboratory Coordinator

Pamela Mason
Laboratory Testing

Julie Hust
Laboratory Testing

W. Stephen Heffron
Laboratory Testing

Acknowledgments

The authors wish to thank the Terrific Science Press Design and Production Team and the following individuals who have contributed to the success of the Teaching Science with TOYS program and to the development of the activities in this book.

Reviewers

Frank Cardulla, Niles North High School, Skokie, IL
Lynn Carlson, University of Wisconsin–Parkside, Kenosha, WI
Baird Lloyd, Miami University, Middletown, OH
Diane Rose, Ursuline Academy, Cincinnati, OH

Contributors to Pedagogical Strategies

Krista Gerhardt, Wilbur Wright Middle School, Dayton, OH; Teaching Science with TOYS, 1993.
Kathy LaRoe, Radley Middle School, East Helena, MT; Teaching Science with TOYS, 1993.
Alyson Mike, Radley Middle School, East Helena, MT; Teaching Science with TOYS, 1993.
Tom Runyan, Middletown High School, Middletown, OH; Teaching Science with TOYS, peer mentor.
Karen Scott, St. Frances DeSales High School, Columbus, OH; Teaching Science with TOYS, 1993.

Teachers

The activities in this and other Teaching Science with TOYS books have been contributed and tested by teachers in the Teaching Science with TOYS program. We wish to acknowledge their efforts in making these activities effective and relevant teaching tools.

University and District Affiliates

Matt Arthur, Ashland University, Ashland, OH
Zexia Barnes, Morehead State University, Morehead, KY
Sue Anne Berger and John Trefny, Colorado School of Mines, Golden, CO
J. Hoyt Bowers, Wayland Baptist University, Plainview, TX
Joanne Bowers, Plainview High School, Plainview, TX
Herb Bryce, Seattle Central Community College, Seattle, WA
David Christensen, The University of Northern Iowa, Cedar Falls, IA
Laura Daly, Texas Christian University, Fort Worth, TX
Mary Beth Dove, Butler Elementary School, Butler, OH
Dianne Epp, East High School, Lincoln, NE
Wendy Fleischman, Alaska Pacific University, Anchorage, AK
Babu George, Sacred Heart University, Fairfield, CT
James Golen, University of Massachusetts, North Dartmouth, MA
Richard Hansgen, Bluffton College, Bluffton, OH
Ann Hoffelder, Cumberland College, Williamsburg, KY
Cindy Johnston, Lebanon Valley College of Pennsylvania, Annville, PA
Teresa Kokoski, University of New Mexico, Albuquerque, NM
Karen Levitt, University of Pittsburgh, Pittsburgh, PA
Maria Galvez Martin, Ohio State University–Lima, Lima, OH
Donald Murad and Charlene Czerniak, University of Toledo, Perrysburg, OH

Hasker Nelson, African-American Math Science Coalition, Cincinnati, OH
Judy Ng, James Madison High School, Vienna, VA
Larry Peck, Texas A & M University, College Station, TX
Carol Stearns, Princeton University, Princeton, NJ
Victoria Swenson, Grand Valley State University, Allendale, MI
Leon Venable, Agnes Scott College, Decatur, GA
Doris Warren, Houston Baptist University, Houston, TX
Richard Willis, Kennebunk High School, Kennebunk, ME
Steven Wright, University of Wisconsin–Stevens Point, Stevens Point, WI

Foreword

Matter is an important part of our world. In fact, everything around us is classified as matter. By studying matter and the changes it can undergo, students can begin to make sense of and to take control of their world. Science provides opportunities for students to do just this. Science allows students to ask questions about the world and to try to find the answers. Students of science should not simply memorize definitions and parrot facts. Rather, students should be engaged in making connections between scientific phenomena and their world.

Investigating Solids, Liquids, and Gases with TOYS is one of several books to result from the National Science Foundation-funded Teaching Science with TOYS programs located at Miami University in Ohio and other affiliated universities throughout the country. The goal of these programs is to enhance teachers' knowledge of chemistry and physics and to encourage the use of activity-based, discovery-oriented science instruction. A key feature of the TOYS programs has been that they promote the use of toys, other common play materials, and household items as ideal scientific tools. From TOYS participants and follow-up evaluations, we have learned that students are receptive to the user-friendly nature of toys and these other items. Also, because these items are a part of the students' everyday world, they help students to appreciate that science is all around them, not just part of a science class in school.

This teacher resource module provides teachers with a basis for engaging their students in the study of the states of matter and changes of state. Aimed at middle-school teachers, this book helps teachers provide opportunities for their students to explore matter in its three states, to compare and contrast the properties between these states, and to determine the conditions necessary for matter to change from one state to another. The book includes a Content Review written for the teachers, a Pedagogical Strategies section that includes ideas generated by teachers who have participated in one of the Teaching Science with TOYS programs, and a collection of 24 physical science activities for teachers to use with their students. These activities have been extensively tested both in TOYS workshops and in the classrooms of TOYS teachers.

The authors of this book and all of the program staff wish to thank all of the TOYS teachers from around the country for the valuable contributions they have made to this effort. We also invite those who are using TOYS materials for the first time to discover the power of teaching students to discover the joy of learning science through this innovative approach.

Mickey Sarquis, Principal Investigator
Teaching Science with TOYS

Investigating Solids, Liquids, and Gases with TOYS

Introduction

This section is an introduction to the TOYS program and this Teaching Science with TOYS Resource Module and its organization. This module focuses on states of matter and changes of state.

WHAT IS TEACHING SCIENCE WITH TOYS?

Teaching Science with TOYS is a National Science Foundation-funded project located at Miami University in Ohio. The goal of the project is to increase teachers' comfort with and understanding of scientific phenomena in our world and to encourage activity-based, discovery-oriented science instruction. As teachers become more knowledgeable about science, they become more confident at nurturing their own and their students' innate sense of wonder and natural curiosity. The TOYS project promotes toys and common play materials as an ideal mechanism for science instruction because they are an everyday part of the students' world and carry a user-friendly message. Through TOYS and its affiliated programs, thousands of K–12 teachers nationwide have brought toy-based science into their classrooms using teacher-tested TOYS activities.

Through written materials such as this TOYS Teacher Resource Module, many more teachers and students can share in the fun and learning of the TOYS project.

To find out more about TOYS programming, contact us at

Center for Chemistry Education
4200 East University Blvd.
Middletown, OH 45042
513/727-3421
FAX: 513/727-3328
cce@muohio.edu
www.terrificscience.org

WHAT ARE TOYS TEACHER RESOURCE MODULES?

TOYS Teacher Resource Modules are collections of TOYS activities grouped around a topic or theme along with supporting science content and pedagogical materials. Each module is prepared for a specific grade range. The modules have been developed especially for teachers who want to use toy-based physical science activities in the classroom but who may not have been able to attend a TOYS workshop at the Miami site or one of the affiliate sites nationwide. The modules do not assume any prior knowledge of physical science—complete content review and activity explanations are included.

The topic of this module is states of matter and changes of state. This module has been developed for use by middle-school teachers but can be modified for use by teachers of younger or older students.

INSIDE THIS TEACHER RESOURCE MODULE

This module is organized into three main sections: Content Review, Pedagogical Strategies, and Activities. We suggest that you skim the Content Review section first to get a feeling for the topic, read the Pedagogical Strategies section to get an overview of the unit, then read each of the Activities to understand them in detail. After that, we recommend that you reread the Content Review section more closely. The following paragraphs provide a brief overview of these sections.

Content Review

The Content Review section is intended to provide you, the teacher, with an introduction to (or a review of) states of matter. The material in this section (and in the individual activity explanations) is designed to provide you with information at a level beyond what you will present to your students. You can then evaluate how to adjust the content presentation for your own students.

The Content Review section in this module is divided into two main parts: States of Matter and Changes of State. States of Matter includes topics such as the characteristic properties of each state, density, and the effect of temperature and pressure on the state of matter. Changes of State includes such topics as energy changes and transitions between different states.

Pedagogical Strategies

The Pedagogical Strategies section is intended to provide ideas for effectively teaching a unit on states of matter. It suggests ways to incorporate the toy-based activities presented in the module into a series of lessons using a constructivist approach and employing a learning cycle philosophy.

Activities

Each module activity provides complete instructions for conducting the activity in your classroom. These activities have been classroom-tested by teachers like yourself and have been demonstrated to be practical, safe, and effective in the typical middle-school classroom. Each activity provides a photograph of the toy or activity setup and summarizes the following information:

- **Hands-On Activity/ Demonstration/ Learning Center:** The recommended presentation style(s) for each activity is indicated along with a suggested grade-level range.

- **Time Required:** An estimated time for conducting the Procedure is listed. This time estimate is based on feedback from classroom testing, but your time may vary depending on your classroom and teaching style.

- **Key Science Topics:** Targeted key science topics are listed.

- **Student Background:** Background knowledge needed by students prior to doing the activity is listed.

- **National Science Education Standards:** This section describes how the activity meets certain Science as Inquiry and Physical Science Standards.

- **Additional Process Skills:** The activity provides opportunities for students to use these science process skills, which may not be mentioned specifically in the National Science Education Standards.

- **Materials:** Materials are listed for each part of the activity, divided into amounts per class, per group, and per student.

- **Safety and Disposal:** Special safety and/or disposal procedures are listed if required.

- **Getting Ready:** Information is provided in Getting Ready when preparation is needed before beginning the activity with the students.

- **Procedure:** The steps in the Procedure are directed toward you, the teacher, and include cautions and suggestions where appropriate.

- **Variations and Extensions:** Variations are alternative methods for doing the Procedure. Extensions are methods for furthering student understanding of topics.

- **Explanation:** The Explanation is written to you, the teacher, and is intended to be modified for students.

- Assessment: Assessment contains strategies for assessing how well students have understood the concepts presented in the activity.

- Cross-Curricular Integration: Cross-Curricular Integration provides suggestions for integrating the science activity with other areas of the curriculum.

- References: References used to write the activity are listed.

- Contributors: Individuals, including Teaching Science with TOYS graduates, who contributed to the development of the activity are listed.

- Handout Masters: Masters for data sheets, observation sheets, and other handouts are provided for some activities.

Notes and safety cautions are included in activities as needed and are indicated by the following icons and type style:

Notes are preceded by an arrow and appear in italics.

Cautions are preceded by an exclamation point and appear in italics.

Activity List

The following activities explore the properties of matter and tie together the concepts involved in changes of state.

STATES OF MATTER

1. **Properties of Matter.** Students investigate the compressibility of solids, liquids, and gases and the effects of temperature changes.

2. **BedBugs.** A BedBugs® game is used to illustrate the motion of particles in different phases; as different amounts of energy are applied in the game, the motion of the bugs changes.

3. **Mystery Eggs.** Students observe the properties of solids, liquids, and gases that are enclosed in rigid containers (plastic eggs). Students then use what they have learned to distinguish between a raw egg and a hard-boiled egg without cracking the shell.

4. **Balloon in a Bottle.** Students attempt to inflate a balloon that is inside a bottle, and thus they observe that a gas (air) takes up space.

5. **Burping Bottle.** Students make a bottle "burp" by pouring water into it and observe that gas (air) takes up space.

6. **Tissue in a Cup.** Students use a plastic cup to keep a tissue dry underwater, and they learn that a gas (air) takes up space.

7. **Showing That Air Has Mass.** Students measure the mass of a soccer ball before and after inflating it and thus observe that air has mass.

8. **Marshmallow in a Syringe.** Students use a marshmallow and a syringe (without the needle) to investigate the relationship between pressure and volume of a gas.

9. **Moving Molecules.** Students observe that the rate at which fluids diffuse is influenced by temperature.

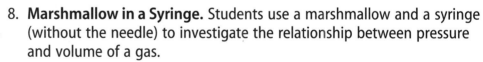

10. **Non-Newtonian Fluids—Liquids or Solids?** Students make and investigate some substances whose properties are hard to classify as solid or liquid.

11. **Rock Candy Crystals.** By making rock candy, students discover that heating the sugar solution enables more sugar to be dissolved in the water. They also observe that when the water is allowed to evaporate from the saturated sugar solution, the sugar crystallizes into rock candy.

12. **Crystals from Solutions.** Students make colorful crystal "trees" by making water solutions and allowing the water to evaporate.

13. **Crystals by Freezing.** Students investigate freezing by experimenting with ice cream and freezer pops.

14. **Boiling Water in a Paper Pot.** Students observe that water can be boiled in paper with an open flame and that a balloon containing a small amount of water will not burst when heated with a match.

15. **Boiling Liquids in a Syringe.** Students make water and/or other liquids boil below their normal (atmospheric pressure) boiling points by lowering the pressure inside a syringe.

16. **Boiling Water with Ice.** This demonstration is somewhat counterintuitive; water is made to boil using ice. It illustrates the effect of pressure on the boiling point of water.

17. **Liquid to Gas in a Flick.** The butane in a lighter is a good example of a substance being changed from the liquid to the gas state as a result of a pressure change.

18. **Disappearing Air Freshener.** A common household air freshener is used to illustrate the process of evaporation.

19. **A Cool Phase Change.** Students observe that some liquids evaporate more readily than others and that evaporation is a cooling process.

20. **Using Dry Ice to Inflate a Balloon.** Students observe a balloon inflate as a small amount of powdered dry ice inside sublimes.

21. **The Phase Changes of Carbon Dioxide.** Students observe the effect of pressure on the phase of dry ice; it sublimes under normal room conditions but melts at higher pressure.

22. **Balloon-into-a-Flask Challenge.** Students make use of the difference in the volumes of gaseous and liquid water to push a balloon into a flask.

23. **Crushing an Aluminum Can.** This activity shows the difference in the volume of water when it is in the gas state and the liquid state.

24. **Hats Off to the Drinking Bird.** Students form hypotheses as to why the bird bobs and then test these hypotheses. This activity relies on student understanding of the changes of state and focuses on the energy changes that accompany these changes of state.

Safety Procedures

Experiments, demonstrations, and hands-on activities add relevance, fun, and excitement to science education at any level. However, even the simplest activity can become dangerous when the proper safety precautions are ignored or when the activity is done incorrectly or performed by students without proper supervision. While the activities in this book include cautions, warnings, and safety reminders from sources believed to be reliable, and while the text has been extensively reviewed, it is your responsibility to develop and follow procedures for the safe execution of any activity you choose to do. You are also responsible for the safe handling, use, and disposal of chemicals in accordance with local and state regulations and requirements.

SAFETY FIRST

- Collect and read the Materials Safety Data Sheets (MSDS) for all of the chemicals used in your experiments. MSDSs provide physical property data, toxicity information, and handling and disposal specifications for chemicals. They can be obtained upon request from manufacturers and distributors of these chemicals or from a world wide web site such as *http://hazard.com/msds/*. In fact, MSDSs are often shipped with chemicals when they are ordered. These should be collected and made available to students, faculty, or parents for information about the specific chemicals used in these activities.

- Read and follow the American Chemical Society Minimum Safety Guidelines for Chemical Demonstrations on page 9. Remember that you are a role model for your students—your attention to safety will help them develop good safety habits while assuring that everyone has fun with these activities.

- Read each activity carefully and observe all safety precautions and disposal procedures. Determine and follow all local and state regulations and requirements.

- Never attempt an activity if you are unfamiliar or uncomfortable with the procedures or materials involved. Consult a high school or college chemistry teacher or an industrial chemist for advice or ask that person to perform the activity for your class. These people are often delighted to help.

- Always practice activities yourself before using them with your class. This is the only way to become thoroughly familiar with an activity, and familiarity will help prevent potentially hazardous (or merely embarrassing) mishaps. In addition, you may find variations that will make the activity more meaningful to your students.

- Do not do activities at grade levels beyond those recommended without careful consideration of safety, classroom management, and the need for additional adult supervision.

- You, your assistants, and any students participating in the preparation for or performance of the activity must wear safety goggles if indicated in the activity and at any other time you deem necessary.

- Special safety instructions are not given for everyday classroom materials being used in a typical manner. Use common sense when working with hot, sharp, or breakable objects. Keep tables or desks covered to avoid stains. Keep spills cleaned up to avoid falls.

- When an activity requires students to smell a substance, instruct them to smell the substance as follows: Hold the container approximately 6 inches from the nose and, using the free hand, gently waft the air above the open container toward the nose. Never smell an unknown substance by placing it directly under the nose. (See figure.)

Use your free hand to gently fan the vapors from the test tube toward your nose.

Wafting procedure—Carefully fan the air above the open container toward your nose. Avoid hitting the container in the process.

- Caution students never to taste anything made in the laboratory and not to place their fingers in their mouths after handling laboratory chemicals.

ACS MINIMUM SAFETY GUIDELINES FOR CHEMICAL DEMONSTRATIONS

This section outlines safety procedures that Chemical Demonstrators must follow at all times.

1. Know the properties of the chemicals and the chemical reactions involved in all demonstrations presented.

2. Comply with all local rules and regulations.

3. Wear appropriate eye protection for all chemical demonstrations.

4. Warn the members of the audience to cover their ears whenever a loud noise is anticipated.

5. Plan the demonstration so that harmful quantities of noxious gases (e.g., NO_2, SO_2, H_2S) do not enter the local air supply.

6. Provide safety shield protection wherever there is the slightest possibility that a container, its fragments or its contents could be propelled with sufficient force to cause personal injury.

7. Arrange to have a fire extinguisher at hand whenever the slightest possibility for fire exists.

8. Do not taste or encourage spectators to taste any non-food substance.

9. Never use demonstrations in which parts of the human body are placed in danger (such as placing dry ice in the mouth or dipping hands into liquid nitrogen).

10. Do not use "open" containers of volatile, toxic substances (e.g., benzene, CCl_4, CS_2, formaldehyde) without adequate ventilation as provided by fume hoods.

11. Provide written procedure, hazard, and disposal information for each demonstration whenever the audience is encouraged to repeat the demonstration.

12. Arrange for appropriate waste containers for and subsequent disposal of materials harmful to the environment.

Content Review

Before conducting activities from this book in the classroom, you may find it helpful to read this section for a review of topics relating to States of Matter and Changes of State.

STATES OF MATTER

An important characteristic of matter is its physical state, or phase. The common states of matter are described as solid, liquid, and gas. Generally, we can determine the state of matter by observing a substance with the naked eye. This mode of observation is referred to as the macroscopic view. The classification of the state of matter is based on observable characteristic properties shown in Table 1.

Table 1: Observable Characteristic Properties of States of Matter				
State	Shape	Volume	Rigidity, Fluidity	Compressibility
solid	definite shape	definite volume	rigid	very low
liquid	conforms to the shape of its container	definite volume	fluid	very low
gas	conforms to the shape of its container	no definite volume; fills all the space available to it	fluid	very high

To better understand the observable characteristic properties of the different states of matter, we must consider the particle nature of matter. Matter is made up of particles: atoms, molecules, or ions. Considering matter on the atomic scale is useful in explaining the differences between the states of matter. (The term "atomic scale" indicates that we are imagining what matter is like at a magnification factor that would allow us to see individual atoms, ions, or molecules.) Figure 1 is a schematic diagram illustrating the relative positions and types of motions of particles in solids, liquids, and gases.

solid　　　　liquid　　　　gas

Figure 1: This schematic diagram illustrates particles in the solid, liquid, and gas states.

Table 2 lists characteristics of the particles on the atomic scale in each state of matter. The particle characteristics listed describe general models for solids, liquids, and gases. Although they are simplified models, they are adequate for predicting many properties of matter.

Table 2: Characteristics of Particles	
State of Matter	Particle Characteristics
solid	particles touching and vibrating but remaining in a fixed position relative to neighboring particles
liquid	particles touching but able to move past one another; not in a fixed position relative to neighboring particles
gas	particles not touching; movement of one particle independent of movement of all other particles; random movement

Solids

As described in Table 2, particles in solids, liquids, and gases interact differently with neighboring particles. The particles in most solids are arranged in an ordered three-dimensional pattern called a crystal structure. For example, table salt (sodium chloride, NaCl) consists of sodium ions and chloride ions. The formula NaCl tells us the ratio of the two types of ions in sodium chloride—one sodium ion for each chloride ion. NaCl has a characteristic crystal structure in which each sodium ion is surrounded by six chloride ions and each chloride ion is surrounded by six sodium ions. Figure 2 is a representation of the crystal structure of sodium chloride. Figure 2a shows only the relative positions of the sodium and chloride ions; 2b shows the ions drawn to scale with respect to size. Both represent the smallest unit of the repeating pattern, called a unit cell. One crystal of table salt contains uncountable numbers of sodium chloride unit cells stacked together in three dimensions. As shown in Figure 2b, the ions in sodium chloride are packed touching one another (as are particles of solids in general). Thus it should not be surprising that solid sodium chloride is not compressible to any large extent.

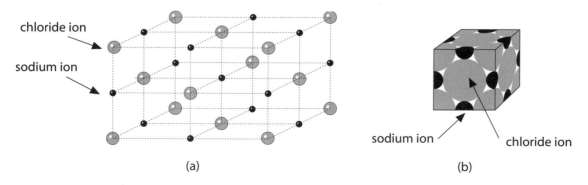

(a)

(b)

Figure 2: (a) Sodium chloride is arranged in crystals—smaller spheres represent sodium ions and larger spheres represent chloride ions. (b) Sodium ions are smaller than chloride ions.

In a crystal, the particles do not change position. This accounts for the rigidity of a solid. However, even in a solid, the particles exhibit motion. But their motion is more restricted in the sense that particles in a solid can only vibrate back and forth.

Like ions, atoms (such as iron) and molecules (such as water) can also exist in the crystalline state. Just as with ions, these particles form regular patterns. Some substances form covalent network solids in which the atoms are held together by covalent bonds. Diamond, which consists solely of carbon atoms bonded to adjacent carbon atoms, is an example of a covalent network solid. (See Figure 3.)

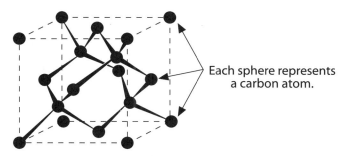

Each sphere represents a carbon atom.

Figure 3: The crystal structure of a diamond.

Table sugar consists of molecules of sucrose in a crystalline structure. If you observe granulated sugar or a sugar cube, you can see the crystals of sugar. However, if you look at a spoonful of powdered sugar, you can no longer see the sugar crystals with your naked eye. Observation at the macroscopic level does not always allow us to predict the arrangement of particles on the atomic scale. In the case of powdered sugar, techniques other than macroscopic observation are necessary to confirm its crystalline nature. For example, X-ray analysis would show that powdered sugar is crystalline.

So far, we have been discussing crystalline solids; most solids fit into this category. Classifying substances allows us to better understand our observations, makes it simpler to see relationships between substances, and aids us in making predictions. If we know a material is a crystalline solid, we can make predictions about its structure and properties. However, when trying to classify, we often come across cases that don't fit neatly into one category or another. Some solids do not form a regular crystalline pattern. These solids are classified as amorphous solids. Candle wax and many plastics are examples of amorphous solids.

The wide range of properties exhibited by plastics is due in part to whether the plastic is crystalline, is amorphous, or has a mixture of crystalline and amorphous regions. Altering the process by which a plastic is made can alter its crystalline content and thus change its properties. For example, both plastic food-storage bags and plastic milk jugs are made of polyethylene. Plastic food-storage bags are made from low-density polyethylene (LDPE, recycle code 4) and have little crystalline content. Milk jugs are made from high-density polyethylene (HDPE, recycle code 2) and are more crystalline. This difference in crystallinity is responsible for the difference in rigidity of the two materials.

Liquids

If we could look at individual particles of a liquid, we would see that the adjacent particles are touching each other and are in motion. Like solids, liquids are not compressible to any large extent because the particles are closely packed. However, the freedom of motion of the particles is greater than in a solid, and a given particle can change position by "sliding" past neighboring particles. This motion of the particles allows a liquid to flow (behave as a fluid) and is the reason that liquids assume the shape of their container. (See Figure 1.)

Gases

Like liquids, gases are fluid—but the particles are very far apart and not touching. The behavior of one gas particle is essentially independent of the adjacent particles. Gas particles are in constant, random motion. They are confined to a location only if they are in a container that is closed. Because gas particles are far apart from each other, they are very compressible. (See Figure 4.)

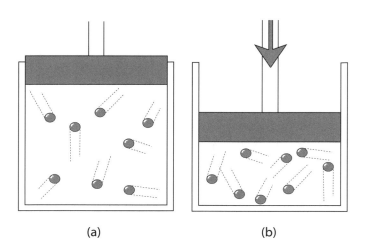

(a)   (b)

Figure 4: (a) Gas particles are far apart and in constant motion. (b) As the gas is compressed, the particles are forced closer together.

Hard-to-Classify Matter

Some matter does not fit our models for solids, liquids, or gases. One example is the liquid crystal. In fact, liquid crystals are considered a fourth state of matter by some scientists. (Plasma is also often referred to as a state of matter but will not be discussed here since a plasma can exist only under extremely high-temperature conditions.) The liquid crystal state has some characteristics of crystalline solids and of liquids. Liquid-crystal displays (LCD) are used on hand-held calculators. Explanations of a liquid-crystal display are given in the Brown and Oxtoby references listed at the end of the Content Review.

Another example of hard-to-classify matter is a mixture of water and corn starch. If held in your hand, the material drips, as we would expect from a liquid. Also, if you place a more-dense object on the surface of the mixture, the object will sink. But if you slap the surface with your hand, the splattering that we would expect from a liquid does not occur. The mixture of water and corn starch is one example of a non-Newtonian fluid. Most fluids are Newtonian fluids; that is, their ability to be easily poured depends on temperature. For non-Newtonian fluids, variables besides temperature affect the fluid's ability to be poured.

Density
Density is a measure of the mass-volume ratio of a substance. At the atomic level, density is a measure of how much matter is packed in a given volume. This will depend upon both the mass and size of the individual particles involved as well as how tightly they are packed together. The particle model of solids, liquids, and gases can help explain the relative densities of these states of matter. Gases have a very low density because the particles are far apart from one another. (As an example, under normal room conditions of temperature and pressure, neon atoms occupy only 0.004% of the space "occupied" by neon gas.) Compared to liquids and solids, a tremendous amount of space exists between particles. Liquids and solids, with particles touching at the atomic level, are much more dense than gases. However, the relative densities of solids and liquids are harder to predict. Logically, the rigid nature of solids implies closer packing and increased density compared to a liquid. In general, this is true—when comparing the solid and liquid states of the same substance, the solids are found to be more dense. For example, imagine heating a solid fat (such as Crisco®) or paraffin in a small saucepan. As the temperature increases, the fat or paraffin melts and begins to take the shape of the container. As the pan fills with liquid, the unmelted solid remains on the bottom of the pan until it is all melted and only the liquid state is present.

Very few substances are less dense in the solid state than in the liquid state. Water is one example. Water is less dense as a solid because liquid water expands in volume when it freezes into ice. Since the same mass has a larger volume, the density of ice is less than that of liquid water. (See Figure 5.)

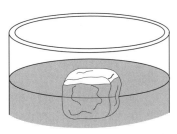

Figure 5: Unlike most substances, the solid state of water is less dense than the liquid state. For this reason, an ice cube will float in a glass of liquid water.

Effects of Temperature and Pressure

Both pressure and temperature affect the state of matter. Therefore, in order to discuss the state in which a substance exists, we must specify a specific temperature and pressure. In the examples used up to this point, the "normal" conditions are room temperature and atmospheric pressure at sea level. We are accustomed to various materials existing in particular states of matter under the normal conditions of our daily lives. For example, we are accustomed to oxygen as a gas, motor oil as a liquid, and iron as a solid.

Materials can exist in different states depending on the temperature. For example, water can be observed in the three states over the temperature range 0°C to 100°C at 1 atm pressure.

Pressure also affects the state of matter. Many gases can be liquefied by increasing the pressure, even if the temperature is well above its normal boiling point. At the molecular level, increasing the pressure forces the gas molecules closer and closer together. When molecules get close enough, their tendency to behave independently is overcome by the strength of the intermolecular attractions, and the gas condenses into the liquid state. For example, butane is a gas at room temperature and 1 atm pressure, but if the pressure is increased, butane can also exist as a liquid at room temperature. Some butane lighters have transparent cases that show butane in the liquid state inside the case. However, when the valve is opened, a gas comes out, not a liquid. (See the activity "Liquid to Gas in a Flick.") Propane is another example of the effect of pressure on the state of matter. You can hear the liquid "sloshing" about in a cylinder of propane indicating it is a liquid inside the cylinder, but a gas comes out when the cylinder valve is opened.

CHANGES OF STATE

A change of state occurs when a substance is transformed from one state of matter to another. Such a change is sometimes referred to as a phase change. In general, any of the three states of matter can change into another state. (See Figure 6.) Table 3 summarizes these changes and gives examples of each type.

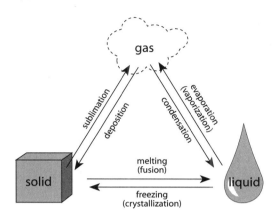

Figure 6: Matter can change from one state to another.

Table 3: Names for and Examples of Changes of State		
Change of State	Name	Examples
solid to liquid	melting, fusion	melting of ice
solid to gas	sublimation	sublimation of dry ice (carbon dioxide) or mothballs (naphthalene or paradichlorobenzene)
liquid to solid	freezing, crystallization	freezing water
liquid to gas	evaporation, vaporization	evaporation of water
gas to solid	deposition	formation of frost and snow
gas to liquid	condensation	formation of dew

Liquid-to-Gas Transitions

When water is placed in an open container, it evaporates. The rate at which it evaporates depends on its temperature. When a liquid is placed in a closed container, some of the liquid will evaporate, but not all of it. What is the difference in these two scenarios? In order to understand this process, let's look at what occurs at the molecular level.

Open Container: The molecules of liquid water are attracted to one another and in close contact. At a given temperature, the average energy of the water molecules is fixed, but individual molecules can have more or less energy than the average. A "high-energy" water molecule on the surface of the water can escape from the rest of the water molecules in the liquid state and become a water molecule in the gas state—water vapor. This process is called evaporation. When the temperature is higher, the average energy of water molecules is higher, and a larger number will have sufficient energy to escape from the rest of the water molecules. What happens to the water that remains in the liquid state? The water molecules that remain in the liquid state are the lower-energy molecules. Because the higher-energy molecules escape, the molecules remaining behind have a lower average energy, and, as a result, the liquid water remaining is cooler than it was before evaporation occurred. In hot weather, we sweat; the result is the evaporation of water which causes cooling. Evaporation also explains why we feel cold when getting out of a swimming pool. As the higher-energy water molecules evaporate, the average energy of the molecules left is lower, so their temperature is also lower.

Closed Container: When an empty container is partially filled with water and sealed, the water begins to evaporate just as it does in an open container. Over a period of time, the level of liquid water decreases slowly, but after a while, there appears to be no more water evaporating and the liquid level remains constant. The sealed container has reached equilibrium, a point at which it appears at the macroscopic level that evaporation has stopped.

What is happening at the molecular level? Some water molecules escape from the liquid water just as they did in an open container. In a closed container, these water vapor molecules are confined to the space within the container. When molecules collide with the walls of the container they exert a force which results in the pressure referred to as the water vapor pressure. As more and more water molecules vaporize, it becomes more likely that some will collide with the water molecules on the surface of the liquid and become part of the bulk liquid again. This process is condensation, the reverse of vaporization. When the water is first placed in the closed container, condensation does not occur because there are no gaseous water molecules. But as evaporation occurs, gas molecules go into the gas state and some gas molecules condense back into the liquid state.

At a given temperature, the rate of evaporation is constant. In the closed container, the rate of condensation increases as the number of gas molecules increases. At some point (assuming there is enough liquid water), the rate of condensation will equal the rate of evaporation. This is described as a dynamic equilibrium. Equilibrium means that the amounts of gas and liquid remain constant. Dynamic means that there is still activity at the molecular level; some liquid molecules are vaporizing and some gas molecules are condensing, but both processes are occurring at exactly the same rate. (See Figure 7.)

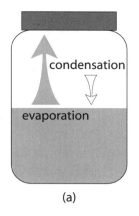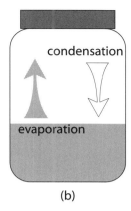

(a)   (b)

Figure 7: (a) Initially, the rate of evaporation is greater than the rate of condensation. (b) At equilibrium, the rates are equal.

When a closed system containing both liquid and vapor has reached equilibrium, the pressure of the vapor is called the equilibrium vapor pressure of the liquid. The equilibrium vapor pressure depends on the temperature of the liquid. As the temperature increases, the equilibrium vapor pressure increases. At the molecular level, this means that at a higher temperature, the average energy of the molecules is higher and a larger fraction of the liquid molecules have sufficient energy to escape from the liquid state. Figure 8 is a graph of vapor pressure versus temperature for water. If the temperature is increased enough, boiling occurs.

Figure 8: The vapor pressure of water depends on temperature.

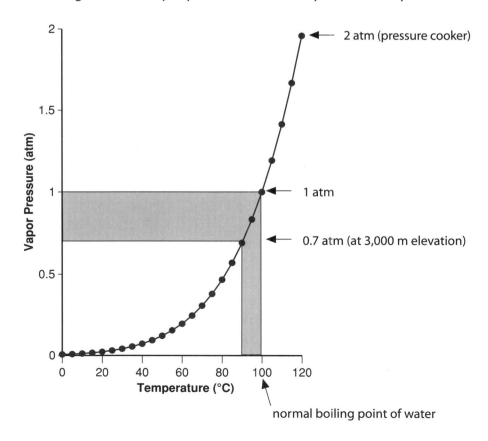

Boiling: Boiling occurs when vapor forms within the liquid. On the macroscopic level, we notice bubbles of gas forming in the liquid and rising to the surface. Boiling occurs when the vapor pressure of the liquid is equal to the confining pressure of the environment.

The temperature at which boiling occurs is called the boiling point. We often say that water boils at 100°C. However, we should say that the normal boiling point of water is 100°C. "Normal" in this context refers to standard pressure, 1.00 atmosphere (760 mm Hg or 760 torr). Usually, when people discuss the boiling point, they mean at 1 atmosphere (atm) pressure.

Can boiling occur at different temperatures? In the mountains, at an altitude of 3,000 m, the boiling point of water is only about 90°C. This is because the atmospheric pressure is lower than 1 atm, so the vapor pressure of water equals the atmospheric pressure at a lower temperature. The atmospheric pressure at 3,000 m is about 0.7 atm. In Figure 8, the corresponding boiling point is seen to be about 90°C. Likewise, at pressures above 1 atm, the boiling point of water is above 100°C. Home canners take advantage of this property when preparing canned food. A temperature of 140°C is required to ensure that certain bacteria are killed. Cooking food in boiling water at atmospheric pressure is not adequate. No matter how long the food is cooked in boiling water at 1 atm pressure, the cooking temperature will not exceed 100°C. Canners use pressure cookers to process the food before canning. The pressure cooker is a closed system that causes the pressure to increase above normal pressure. As a result, the water in which the food is cooked can reach a temperature higher than 100°C before it boils. Use Figure 8 to estimate the pressure necessary to cause water to boil at 110°C.

Gas-to-Liquid Transitions

Condensation is the reverse of evaporation. If the vapor pressure of a gas is higher than the equilibrium vapor pressure of the liquid, gas will condense and form liquid until equilibrium is established. With water, we can see this occur in a variety of ways. On a cool morning, fog forms. Fog consists of water droplets so small that they remain dispersed in air rather than falling to the ground as rain. The water droplets have formed due to condensation. If the air is saturated with water vapor at a particular temperature, cooling will cause condensation to occur. The same phenomenon can occur when a person wearing glasses walks into a warm room from the cold outdoors. The vapor pressure of water in the warm room is higher than the vapor pressure of water at the colder outdoor temperature. So water vapor cooled by contact with the colder glasses condenses to liquid water on the surface of the glasses. Other examples of water vapor condensation include water forming on the outside surface of a glass containing ice water or on the surface of a can or jar removed from a refrigerator or freezer.

Solid-to-Liquid Transitions

At room temperature, an ice cube melts to form liquid water. But an ice cube in a freezer remains solid. The determining factor is the temperature. At temperatures below 0°C, water is a solid, and at temperatures above 0°C (but below 100°C), water is a liquid. The transition from solid to liquid is called melting. As with the liquid-to-gas transition, the temperature at which water melts depends on the pressure. However, the change of the melting point due to changes in atmospheric pressure is small enough that it can be ignored.

At the molecular level, we can imagine the molecules in the ice crystal vibrating but not moving relative to adjacent molecules. As the temperature increases, the vibration increases; and at the melting point, the energy is sufficient to cause the structure of ice to collapse, and the water is changed into the liquid state. Molecules are still in close proximity with their neighbors, but they are able to change neighbors.

Liquid-to-Solid Transitions

Water in an ice cube tray changes from the liquid state to the solid state over a period of time when placed in a freezer. Cooling removes energy from the water, and the motion of liquid molecules is slowed. At 0°C, the water begins to freeze and the motion of the molecules becomes restricted as they assume the structure of solid water.

Solid-to-Gas Transitions

The most commonly cited example of a substance that sublimes is dry ice, solid carbon dioxide. At room temperature, dry ice sublimes, which means it goes directly from the solid state to the gas state. While this behavior under normal conditions is not the usual case for most solids, it is by no means unique. The solid crystals in moth balls gradually sublime and the resulting vapor is responsible for the characteristic odors associated with moth balls. Even ice will sublime. Because of sublimation, an ice cube that remains in a freezer for a long period of time will gradually get smaller, even though the temperature in the freezer never rises above the freezing point. Likewise, snow and ice can slowly "disappear" even if the temperature stays below 0°C.

Gas-to-Solid Transitions

The formation of frost during winter or on the inside of a freezer is due to the reverse of sublimation, called deposition. It is water vapor inside the freezer that deposits as frost inside the freezer. If it first condensed to a liquid and then froze, we would expect the buildup of frost to occur only on the bottom of the freezer, since liquid water is more dense than air. But frost forms on all surfaces, indicating that water vapor deposition has occurred.

Energy Changes

Energy changes accompany changes of state. Table 4 lists the name(s) for each change of state. The symbol ΔH represents the change in heat content that accompanies the transition. A positive sign indicates that heat is added to the substance; that is, heat is absorbed from the surroundings. Changes of state with a positive ΔH are said to be endothermic. A negative ΔH means that heat is lost from the substance; that is, heat is released to the surroundings. This is an exothermic change.

Table 4: Changes of State and Accompanying Energy Changes		
Change of State	Energy Change	ΔH
solid to liquid (fusion, melting)	heat absorbed, endothermic	+
liquid to solid (crystallization, freezing)	heat released, exothermic	−
liquid to gas (vaporization, evaporation)	heat absorbed, endothermic	+
gas to liquid (condensation)	heat released, exothermic	−
solid to gas (sublimation)	heat absorbed, endothermic	+
gas to solid (deposition)	heat released, exothermic	−

To convert a liquid to a gas at its boiling point, energy must be added; this energy is called the heat of vaporization. The reverse process, condensation, releases energy to the environment. The energy released is called the heat of condensation. The amount of heat transferred in both cases is exactly the same; only the direction of the transfer is different. The same is true for the other pairs: heats of fusion and crystallization and heats of sublimation and deposition.

Compared to other liquids, a lot of energy is required to boil water because water molecules have unusually strong forces of attraction for each other, including hydrogen bonding as shown in Figure 9. These forces must be overcome before water molecules can escape to the gas phase where they are independent of the other water molecules. Likewise, a large amount of energy is released when steam (gaseous water) condenses. This is why steam burns are potentially much more serious than burns from hot water.

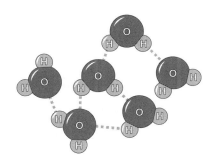

Figure 9: Hydrogen bonding (dotted lines) causes water molecules to be highly attracted to one another.

The value of the heat of fusion for ice is 334 J/g; that is, 334 Joules of heat are needed to change 1 gram of solid ice at 0°C to liquid water at 0°C. The corresponding value for the heat of vaporization is 2,260 J/g. (In calories, the corresponding values for heat of fusion and vaporization are 80 cal/g and 540 cal/g, respectively.) Note that the heat of vaporization is a much larger value than the heat of fusion. The large value of the heat of vaporization is consistent with the fact that all attractive forces must be overcome as the liquid-to-gas state change occurs. Figure 10 shows a plot of temperature versus time for water. A sample of ice at −20°C is heated at a constant rate. Note that during a phase change, the temperature remains constant. Initially, the temperature of the ice rises, but when it reaches 0°C, the ice begins to melt and the temperature remains constant at 0°C until all the ice has melted. Then the temperature rises again until the water starts to boil. Once again, the temperature remains constant at the boiling point of water until all of the water has been converted to steam. Only when all the water is steam can the temperature of the steam rise. Since the graph in Figure 10 represents heat being added at a constant rate, the lengths of the horizontal lines are indicative of the relative magnitudes of the heats of fusion and vaporization.

Figure 10: Heating curve for water

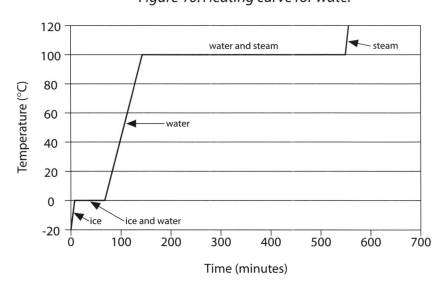

REFERENCES FOR CONTENT REVIEW

Brown, T.L.; LeMay, H.E.; Bursten, B.E. *Chemistry, The Central Science,* 6th ed.; Prentice Hall: Englewood Cliffs, NJ; 1994, pp 415–421.

Carle, M.A.; Sarquis, M.; Nolan, L.M. *Physical Science, The Challenge of Discovery;* D.C. Heath: Lexington, MA, 1991.

Holtzclaw, H.F., Jr.; Robinson, W.R.; Nebergall, W.H. *College Chemistry,* 7th ed.; D.C. Heath: Lexington, MA; 1984, p 698.

Oxtoby, D.W.; Nachtrieb, N.H.; Freeman, W.A. *Chemistry: Science of Change,* 2nd ed.; Saunders College: Philadelphia, 1994; pp 827–829.

Pedagogical Strategies

A variety of instructional approaches can be effectively used to present the science content in this book. To maximize active student learning, we have included specific suggestions for addressing the National Standards for Science Education; for developing science process skills in your students; and for incorporating science journals, cooperative learning, and active assessment in your classroom. These pedagogical strategies are based on modern methods of science education that originated with theories of cognitive functioning introduced by Jean Piaget; these theories are summarized in the following statements:

- If students are to give up misconceptions about science, they must have an opportunity to actively reconstruct their world view based on exploration, interpretation, and organization of new ideas (Bybee, 1990).
- Exploration, interpretation, and organization of new ideas are most effective in a curriculum where hands-on, inquiry-based activities are integrated into learning cycle units (Renner, 1988).
- Hands-on, inquiry-based activities that involve conceptual change, problem solving, divergent thinking, and creativity are particularly effective in cooperative-learning situations (Johnson, 1990).

The suggestions in this section can be applied in a variety of ways. This section is intended to provide you with a basis for integrating the activities in this book into your own curriculum. We encourage you to modify the presentation to meet the needs of your students and fit your own style of teaching.

USING NATIONAL STANDARDS FOR SCIENCE EDUCATION

Many scientists, educators, policy-makers, and parents share a vision of education in America in which *all* students will become literate in science, mathematics, and technology. This vision has been translated into sets of standards for science education, most notably the *National Science Education Standards* compiled under the direction of the National Research Council and the *Benchmarks for Science Literacy* compiled under the direction of the American Association for the Advancement of Science. This book is organized around the Physical Science Content Standard for Grades 5–8 from the *National Science Education Standards.* The activities in this book are consistent with the recommendations of both *Benchmarks for Science Literacy* and the *National Science Education Standards.*

In your day-to-day teaching you probably do not work directly with these standards. Instead, you probably work with the required science curriculum that has been developed for your district. So why do you need to know about the *National Science Education Standards* and *Benchmarks for Science Literacy* and how this book relates to them? You need to know for several reasons:

- Your local curriculum may be based on one or both of these sets of standards.
- You may be part of your district's efforts to develop new science curricula based on one or both of these sets of standards.
- Your current science curriculum may be inadequate, and you may need information to prepare you to encourage your school system to reform the curriculum.
- You may serve on a committee to select science textbooks for your district and need a frame of reference for selecting the best materials.
- You may need to justify the added expense of hands-on, minds-on science education to supervisors, colleagues, and parents.
- You may need to justify the instructional value of the activities in this book to supervisors, colleagues, and parents who are not familiar with activity-oriented science.

In the following sections, we discuss two major categories of standards found in the *National Science Education Standards:* science as inquiry and physical science.

Science as Inquiry

When we think about teaching physical science, we often begin by thinking about content objectives, such as the following: Students will learn that air takes up space, students will learn that dry ice sublimes under normal room conditions, or students will learn the properties of each of three states of matter. But each experience students have with science investigations in the classroom does more than teach science content; it also helps to shape student perceptions of what science is and what scientists do. Thus, every activity in this book has been developed with the goal of shaping students' ideas about science and about how they can work as scientists. The *National Science Education Standards* states that as a result of activities in grades 5–8, students should develop the abilities necessary to do scientific inquiry and to understand the process of scientific inquiry.

Although the Science as Inquiry portion of the National Standards incorporates the concepts of process skills into the statements describing the Abilities Necessary to Do Scientific Inquiry, a list of specific process skills by name (such as "comparing/contrasting") is not provided. In this book, the Science as Inquiry statements encompass elements of many process skills. Where appropriate, we have also included a section called "Additional Process Skills" that names other important process skills.

Physical Science

An understanding of the physical world is an essential component of scientific literacy. Remember that students are expected to develop an understanding of science concepts as a result of observation and manipulation of objects and materials in their environment. Every activity in this book gives students the opportunity to develop their understanding through such experiences.

Relating the Activities in This Book to the National Standards

Every activity in this book can be used to help students develop skills and knowledge specified by the *National Science Education Standards.* To help you in selecting activities and preparing lessons, the first page of each activity lists two main categories of standards and briefly describes how the activity relates to those standards. Also, the National Science Education Standards Matrix Appendix shows which standards are met by each activity.

EFFECTIVELY INCORPORATING THE ACTIVITIES IN THE CLASSROOM

This book focuses on states of matter and changes of state. The activities presented in this book can easily supplement any existing middle-school curriculum on the states of matter or can be used to build your own unit. The activities included are intended to help you engage your students in hands-on science investigations of the properties of solids, liquids, and gases and changes between these states. While the target audience is denoted in each of the activities, the activities can be adapted for use with younger or older students.

Selecting the Activities

The activities in this collection incorporate three different presentation styles: hands-on activities, demonstrations, and learning center activities. Many of the activities can be conducted in more than one of these forms; you can choose whichever form fits your students' needs and abilities. Recommended grade levels for learning centers do not include grades beyond 8, because learning centers are not commonly used at the high school level. However, teachers who choose to use the learning center format for grades 9–12 can adapt activities as necessary.

Which activities you do and the order in which you do them are up to you. Some teachers may choose to concentrate on one state of matter a week, doing several of the activities involving that state during the week; others may prefer a much shorter time frame and select a different state and corresponding activity (or activities) to be done each day.

Using Science Journals

Writing is an integral part of the process of scientific inquiry. When working as scientists, students must record observations and data, organize and summarize results, and draw conclusions. Often they must communicate both the process of the investigation and its results to others through words and/or pictures.

Writing can be viewed as hands-on thinking: it gives students opportunities for reflection and active processing of learning experiences. As a result of regular writing experiences, students gain a better understanding of what they know. Regular use of science journals throughout the lessons in this book will give students many opportunities to practice writing skills in the context of doing science.

Cooperative Learning

You may wish to use some formal cooperative strategies while doing these activities. Cooperative learning has been well documented as enhancing student achievement. Several popular models for cooperative groups have been described by D.W. and R.T. Johnson, Robert Slavin, Spencer Kajan, Eliot Aronsen, and others. Although these models vary, they typically include elements of group goals or positive interdependence and individual accountability. Often, cooperative models suggest specific roles for each student. If students have not been using cooperative learning routinely, time must be spent at the beginning of the assignment explaining individual accountability and group interdependence and reviewing social skills needed for cooperative group work. Students should understand that their grades are dependent upon each person carrying out the assigned task. The teacher should observe the groups at work and assist them with the necessary social and academic skills.

ASSESSMENT

Assessment and learning are two sides of the same coin. Assessments enable students to let teachers know what they are learning; and when students engage in an assessment exercise, they should learn from it. Paper-and-pencil tests are the familiar and prevalent form of assessment. But in light of what we are hoping to teach students about both the process and content of science, traditional tests requiring students to choose one of a few given answers or to fill in the blank measure only a fraction of what we need to know about their science learning. The *National Science Education Standards* advocates using diverse assessment methods, including performances and portfolios as well as paper-and-pencil tests.

Emerging from among a host of terms describing current assessment options (for example, authentic, alternative, portfolio, and performance), the term active assessment has been proposed by George Hein and Sabra Price in their book, *Active Assessment for Active Science*. They define "active assessment" as a whole family of assessment methods that actively engage the learner and can also be interpreted meaningfully by the teacher.

Almost any of the experiences that make up the activities in this book can also serve as active assessments. For example, brainstorming sessions, science journal entries, data and observations from science investigations, and writing extensions can all be part of developing a picture of what students are learning. We hope that as you use the activities in this book, you will engage in many different forms of active assessment, thus maximizing the opportunity for all students to demonstrate their accomplishments and understanding. Several ideas for assessment methods are discussed in this section:

     1. Activity Performance Evaluation
     2. Situation Questions—Pre- and Post-Test
     3. One-Minute Concept Questions
     4. Situation Comparison—Cooperative Group
     5. Activity and Demonstration Practical
     6. Concept Map—Pre- and Post-Map
*   7. Five-Minute Videotape Production
*   8. Concept Poster
*   9. Concept Cartoon
*   10. Concept Song

*   You may elect to assign all students one of these types of evaluation (#7–10) or have them choose one of the four types for variety in the presentations. Some samples of evaluation sheets and student handouts are provided in Appendix A beginning on page 267.

Activity Performance Evaluation

The purpose of an activity performance evaluation is to assess the process skills of the middle-school student. The evaluation should be based on individual performance on a given activity and the results provided as immediate feedback. This performance evaluation can also be used in a cooperative learning environment for assessment of an entire group's performance. The accumulated score should be based on the student (or group) meeting your goals for the lesson.

Situation Questions—Pre- and Post-Test

The purpose of this type of test is to assess the students' knowledge of the concepts and the ability to apply the concepts before the unit and after the unit. Both written questions and pictures (or demonstrations) should be included to address different learning styles. *A sample pre- and post-test are included in Appendix A on page 269.*

One-Minute Concept Questions

The purpose of this type of assessment is to get immediate feedback on students' understanding of the given topic. Give students a question that may also include a demonstration or illustration and allow them to respond for 1 minute. Sample questions are provided.

1. A piece of dry ice placed on a table slowly "disappears." What happened to the molecules in the dry ice? Did a change of state occur? Explain your reasoning.

2. You place water in a tea kettle and set it on a stove burner. You turn the stove burner to "high" and wait until you hear the tea kettle whistle. In which state did the water molecules touch but not always have the same neighbor?

Situation Comparison—Cooperative Group

The purpose of this type of assessment is to reinforce the concepts from the labs and activities. Give the cooperative groups similar situations and have each group discuss their situation and relate it to activities they have done. Group members should come up with a group answer with evidence. Groups can share their situations and answers with the class. The following is a sample situation comparison.

Situation: In Alaska, the temperature is so cold that the glacial ice does not melt. However, evidence shows that the amount of ice decreases over time. You have an idea of what may be happening to the glacial ice, so you set up a small experiment. During very cold weather when the temperature remains below the freezing point of water, you place a small piece of ice in an uncovered cup and set it outside. You monitor it on a regular basis over several days. The ice decreases in size, but there is no evidence of water in the cup. [Assume no animal (or anything else) was able to get to the ice.]

a) What activity have you done that was closely related to this activity? *"Using Dry Ice to Inflate a Balloon."* (Students may answer "Disappearing Air Freshener." If so, remind them that while the air freshener appears to be a solid subliming, in fact liquid perfume is evaporating slowly from a gel.)

b) As a group, decide what happened to the ice and provide evidence to support your decision.

Activity and Demonstration Practical

The purpose of this type of assessment is to evaluate student process skills in conjunction with the science concepts being studied. Set up stations and allot a certain amount of time per station. (Give students a trial run with this type of assessment if they are not familiar with the format.)

Concept Map—Pre- and Post-Map

This type of assessment allows students to show the relationships between various concepts with arrows and words. It would be most effective as a pre-test and post-test to determine the level of understanding following the module. (Give students an example using a different topic if they are not familiar with concept maps.)

Five-Minute Videotape Production

This type of assessment addresses the needs of the creative and visual learner. Each student or group of students should produce a videotape based on a given set of criteria (such as those in the following list). You may want the other students to have input on this evaluation. *A sample evaluation form is provided in Appendix A on page 271.*

Five-Minute Videotape Production Criteria:

1. Produce a video on one topic from this module.

2. Provide an explanation of the concepts involved during the video.

3. If misconceptions are addressed, appropriate discussion to correct the misconceptions must be included.

Concept Poster

This type of assessment addresses the needs of the creative and visual learner. Each student or group of students should either draw or cut out pictures that respond to a given set of criteria (such as those in the following list) and present the poster to the class. You may want the other students to have input on this evaluation. *A sample evaluation form is provided in Appendix A on page 272.*

Concept Poster Criteria:

1. Prepare a poster on one topic from this module.

2. Present the poster to the class, providing an explanation of the concepts involved in each picture.

3. If misconceptions are addressed, include appropriate discussion to correct the misconceptions.

Concept Cartoon

This type of assessment addresses the needs of the creative and visual learner. Each student or group should draw a cartoon based on a given set of criteria (such as those in the following list). You may want the other students to have input on this evaluation. *A sample evaluation form is provided in Appendix A on page 273.*

Concept Cartoon Criteria:

1. Prepare a cartoon on one topic from this module.

2. Present the cartoon to the class, providing an explanation of the concepts involved in the drawings and the story.

3. If misconceptions are addressed, include appropriate discussion to correct the misconceptions.

Concept Song

This type of assessment addresses the needs of the creative learner. Each student or group of students should compose a song based on a given set of criteria (such as those in the following list). You may want the other students to have input on this evaluation. *A sample evaluation form is provided in Appendix A on page 274.*

Concept Song Criteria:

1. Compose a song on one topic from this module.

2. Sing the song for the class and provide an explanation of the concepts involved in the lyrics.

3. If misconceptions are addressed, include appropriate discussion to correct the misconceptions.

REFERENCES FOR PEDAGOGICAL STRATEGIES

Bybee, R.; Lands, N. "Science for Life and Living: An Elementary School Science Program from Biological Science Curriculum Study," *American Biology Teacher. 52*(2), 1990, 92–98.

Johnson, D.; Johnson, R.; Holubec, E. *Cooperation in the Classroom.* Interaction: Edina, MN, 1990.

Renner, J.W.; Abraham, M.R.; Birnie, H.H. "The Necessity of Each Phase of the Learning Cycle in Teaching High School Physics," *Journal of Research in Science Teaching. 25*(1), 1988, 39–58.

States of Matter

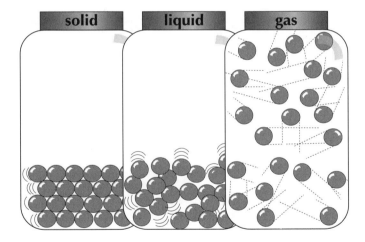

Properties of Matter

...Students use a syringe to explore the concept of compressibility and the effect of temperature on volume for solids, liquids, and gases.

✔ Time Required

Setup	10	minutes
Performance	20	minutes
Cleanup	5	minutes

✔ Key Science Topics

- compressibility
- effect of temperature on volume
- properties of matter

Compressing matter in a syringe

✔ National Science Education Standards

Science as Inquiry Standards:

- Abilities Necessary to Do Scientific Inquiry

  Students make systematic observations and accurate measurements.

  Students use appropriate tools and techniques to gather data.

  Students use data to form a logical argument about the cause-and-effect relationships in the experiment.

Physical Science Standards:

- Properties and Changes of Properties in Matter

  Two characteristic properties of gases are that they are more compressible than solids or liquids and their volume is affected by temperature.

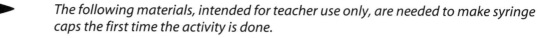

MATERIALS

For Getting Ready only

The following materials, intended for teacher use only, are needed to make syringe caps the first time the activity is done.

Per group
• Bunsen burner or candle
• 2 pairs of forceps, tweezers, or pliers
• small container of water
• 3 disposable syringe needles

Syringe needles are available from Fisher Scientific (#14-826-5B), 1600 W. Glenlake Ave., Itasca, IL 60143; 800/766-7000. Students will NOT be using the needle in this activity; they will use only the plastic connector.

For the Procedure
Per group
• 3 60-mL disposable plastic syringes with Luer-Lok® tips

Syringes are available from Fisher Scientific (#14-823-2D); syringes might also be available from a veterinarian or a farm supply store.

• 3 syringe caps made in Getting Ready
• table salt (sodium chloride, NaCl)
• ice
• containers to hold hot water and ice water
• thermometer
• hot and cold tap water

Generally, hot water from a faucet at about 50–60 °C is hot enough. If hot tap water is unavailable, heat water (no hotter than 70 °C) on a hot plate or by other means.

For the Extensions
❷ All materials listed for the Procedure plus the following:
Per class
• balance

SAFETY AND DISPOSAL

Prepare the syringe cap (see Getting Ready) outside of class. Take care not to get melted plastic on your skin. Proper fire safety should be exercised, such as working on a fire-resistant surface and removing unnecessary flammable materials from the area. Long-haired people should tie hair back when working near flame. Dispose of the needle by placing it in the plastic needle cover and wrapping tape over the open end. Place the wrapped needle cover in the trash.

Because of the potential breakage of thermometers, use metal cooking or alcohol thermometers.

GETTING READY

Take care not to get melted plastic on your skin.

1. Prepare an inexpensive cap for each syringe as follows:

   a. Hold a disposable syringe needle with a pair of pliers, forceps, or tweezers, and use a candle or Bunsen burner to heat the needle close to where it enters the plastic connector.

   b. As the plastic begins to melt, pull the needle out with the second pair of pliers. (See Figure 1.) Drop the hot needle into the small container of water to cool.

   c. After the plastic has cooled, check the cap for leaks as follows: Place the cap on the syringe with the plunger completely in the "in" position. Pull the barrel until it is about halfway out and then release the barrel. If it returns to its original position, the syringe cap is properly sealed. If the barrel does not return to its original position, some air is in the syringe, and the cap leaks. Try heating the plastic tip again to get it to seal or discard the cap and try again.

 If the plastic catches fire, extinguish the fire by dipping the cap in water. Once prepared, the cap may be saved and used again. Dispose of the needle as specified in Safety and Disposal.

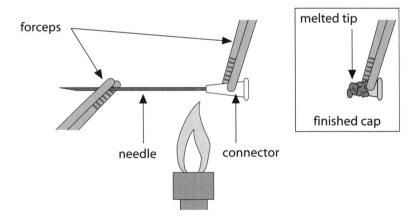

Figure 1: Remove the syringe needle to make a cap.

2. Fill a container with hot water. Generally, hot water from a faucet at about 50–60°C is hot enough. If your tap water is not hot enough or if you desire a more dramatic effect, heat the water (to no higher than 70°C) on a hot plate or by other means.

3. Prepare an ice-water bath by placing ice and water in a container. The temperature should be about 0°C. (If you want a lower temperature, you can add some table salt, about 1 part salt for 10 parts ice.)

INTRODUCING THE ACTIVITY

Show students the matter they are to study: air, water, and salt. Ask what state of matter each material represents. Ask students to share their initial observations of these materials. Divide students into small groups and have them record their observations on the Data Sheet (provided).

Show students the syringe. Ask students what they could observe about each material if it were in a syringe. If no one suggests that syringes are useful for measuring volume and changing pressure, lead students to consider these ideas.

PROCEDURE

1. Show students how to fill and empty the syringe:

   ○ Air—Fill the syringe by pulling the plunger back to about the 20-mL mark and placing the cap on the syringe tip. Empty the syringe by removing the cap and pushing in the plunger.

   ○ Water—With the plunger pushed in to the 0-mL mark, place the syringe tip into a container of water. Pull the plunger back until it is about to the 20-mL mark. Remove the syringe and hold it so the tip is pointed up. Carefully push in the plunger until all air is expelled. Then place the cap on the syringe tip. Empty the syringe by removing the cap and pushing in the plunger to force the water out the tip.

   ○ Salt—Use a dry syringe. With the plunger and cap removed and your finger over the tip of the syringe, pour about 20 mL salt into the barrel of the syringe. Tap the syringe gently so the salt is packed at the bottom. Then place the plunger back into the barrel and turn the syringe so the tip points up. Remove your finger and push in the plunger until as much air as possible is expelled from the barrel. Place the cap on the syringe tip. To remove the salt, remove the cap, turn the syringe so the tip points up, pull out the plunger, and pour out the salt.

2. Tell students that one characteristic difference between states of matter is the compressibility of a substance. Ask, "How can we use these syringes to study the compressibility of our three substances?" Through discussion, lead students to realize that they can put material in the syringe and push in the plunger. Ask, "How can we collect data to measure how compressible each material is?" If no one suggests doing so, lead students to realize that they can record the volume measurements before and after compressing. Ask students if they think the amount of material they start with in the syringe will affect their results. Discuss identifying and controlling variables.

3. Have students work in small groups to investigate the compressibility of the three substances and record their results on the Data Sheet. Tell them to make sure the syringe cap is on securely before trying to push in the plunger.

4. Tell students that another characteristic difference between states of matter is the effect of temperature on volume. Ask students how they can use the syringes to observe the effect of temperature on the volume of their three materials. Lead the students to the conclusion that they can control the temperature by placing syringes in water of different temperatures. Discuss variables and data collection techniques. (Students will need to assume that the syringe contents are originally at room temperature and finally at the temperature of the water the syringe is submerged in.)

5. Have students work in small groups to investigate the temperature-volume relationship for the substance in their syringe and record their results on the Data Sheet. Students should read the temperature of the room and record it as the original temperature of the syringe contents. Record the initial volume. Place the syringe in the hot water. After 5 minutes, read the temperature of the hot water and record it as the final temperature of the syringe contents. Record the final volume.

Tell students that the plunger tends to resist movement due to friction. Tell them to push in gently on the plunger before recording the final volume to make sure the plunger has not become stuck.

6. Have students repeat Step 5 using an ice-water bath.

7. When each group has finished their tasks, compare and contrast the behavior of solids, liquids, and gases with respect to compressibility and temperature-volume relationship. Have students infer reasons for differences in the pressure-volume and temperature-volume relationships of different states of matter.

VARIATIONS AND EXTENSIONS

1. Set up the activity in a learning center and have students conduct the investigation using the Instruction and Observation Sheet (provided).

2. Have students calculate the density of the water after measuring the mass and volume. Ask, "Can the density of salt be calculated in the same way?" (The density would probably not be accurate. In contrast to a liquid, the salt sample contains air which occupies space between adjacent crystals. Thus the density measured is not that of pure salt but of a salt-air combination.)

EXPLANATION

The following explanation is intended for the teacher's information. Modify the explanation for students as required.

One of the characteristic differences between gases and both solids and liquids is the degree of compressibility. While gases are relatively easy to compress, solids and liquids are not. At a molecular level, this difference can be related to the relative proximity of the particles that make up the substance. In gases, the particles have lots of space between them; in solids and liquids, the particles essentially touch each other. Thus gases can be compressed a great deal but solids and liquids cannot be compressed to any measurable extent at the pressures achieved in this activity. (However, students may observe some movement of the syringe barrel as the rubber seal on the syringe barrel becomes deformed when pushed in with water or salt in the syringe. It is also sometimes possible to compact the salt-air combination.)

Another characteristic difference between gases and both solids and liquids is the extent to which temperature changes affect the volume of the sample. When the gas is cooled or warmed, the syringe barrel will move as the volume of the gas changes. Heating gas particles increases their energy, causing them to move more quickly. This increased motion causes them to hit the sides of the container more often, which produces a greater force on a given area and thus a greater pressure. In a syringe with a moveable piston, as the temperature increases, the pressure will first increase. But since only the atmosphere is pushing back, the higher pressure inside the syringe will push the piston out until the pressure inside is once again equal to atmospheric pressure.

Temperature calculations involving gases must use Kelvin temperatures ($K = °C + 273$). The Kelvin scale is an absolute temperature; a temperature of 0 K (-273°C) is the temperature at which the volume of a gas is theoretically zero (and also the temperature at which all molecular motion stops).

The volume of a gas is proportional to its Kelvin temperature. Therefore, if you know a gas's volume at one temperature, you can calculate the gas's volume at a different temperature. A given volume of gas at 0°C (273 K) will increase in volume at 100°C (373 K) by a ratio of 373/273, a factor of 1.37. With solids and liquids, particles remain in close proximity and temperature changes do not affect the volume to anywhere near as large an extent. Thus, the effect of temperature on volume will be much less for solids and liquids, and probably no measurable change will occur under the experimental conditions used here.

ASSESSMENT

Have the students make drawings to represent the arrangement of the particles that make up the solid, the liquid, and the gas. Have students explain why they drew what they did, citing observations from the experiment to support their ideas. Use the students' drawings to diagnose misconceptions about matter. Ask students to evaluate the drawings to determine the best model.

CROSS-CURRICULAR INTEGRATION

Home, safety, and career:

*   Have students study how hydraulic braking systems work and why it is important for such systems to be free of trapped gas bubbles.

HANDOUT MASTERS

Masters for the following handouts are provided:

*   Data Sheet
*   Instruction and Observation Sheet

Copy as needed for classroom use.

Names _____   _____

_____   _____

Date _____

Properties of Matter
Data Sheet

Investigate each substance and record your observations and results of their compressibility.

Substance	State of Matter	Initial Observations	Compressibility	
air			original volume:	
			final volume:	
			change:	
water			original volume:	
			final volume:	
			change:	
salt			original volume:	
			final volume:	
			change:	

Data Sheet, page 2

Investigate the temperature-volume relationship for each substance and record your results.

Substance	Temperature Increase in Hot Water			Effect of Increasing Temperature on Volume			Temperature Decrease in Cold Water			Effect of Decreasing Temperature on Volume		
air	original temperature:	final temperature:	change:	original volume:	volume after heating:	change:	original temperature:	final temperature:	change:	original volume:	volume after cooling:	change:
water	original temperature:	final temperature:	change:	original volume:	volume after heating:	change:	original temperature:	final temperature:	change:	original volume:	volume after cooling:	change:
salt	original temperature:	final temperature:	change:	original volume:	volume after heating:	change:	original temperature:	final temperature:	change:	original volume:	volume after cooling:	change:

Reproduced from *Investigating Solids, Liquids, and Gases with **TOYS***, published by Terrific Science Press.

43

Names _____   _____

          _____   _____

Date   _____

Properties of Matter
Instruction and Observation Sheet

1. Fill and empty the syringe.

    ◦   Air—Pull the plunger back to about the 20-mL mark and place the cap on the syringe tip. Empty the syringe by removing the cap and pushing in the plunger.

    ◦   Water—With the plunger pushed in to the 0-mL mark, place the syringe tip into a container of water. Pull the plunger back until it is about to the 20-mL mark. Remove the syringe and hold it so the tip is pointed up. Carefully push in the plunger until all air is expelled. Then place the cap on the syringe tip. Empty the syringe by removing the cap and pushing in the plunger to force the water out the tip.

    ◦   Salt—Use a dry syringe. With the plunger and cap removed and your finger over the tip of the syringe, pour about 20 mL salt into the barrel of the syringe. Tap the syringe gently so the salt is packed at the bottom. Then place the plunger back into the barrel and turn the syringe so the tip points up. Remove your finger and push in the plunger until as much air as possible is expelled from the barrel. Place the cap on the syringe tip. To remove the salt, remove the cap, turn the syringe so the tip points up, pull out the plunger, and pour out the salt.

2. One characteristic difference between states of matter is the compressibility of a substance. Investigate the compressibility of air, water, and salt. For each substance, put the substance in the syringe and measure and record the original volume. Push in the plunger as far as possible. (Make sure the syringe cap is on securely before trying to push in the plunger.) Measure and record the final volume. Record your results on the Data Sheet. What happens when the syringe plunger is pushed inward?

    _____

    _____

3. Another characteristic difference between states of matter is the effect of temperature on volume. Investigate the temperature-volume relationship for each substance in a syringe. Assume the syringe contents are originally at room temperature. Read the temperature of the room and record it on the Data Sheet as the original temperature of the syringe contents. Record the initial volume. Place the syringe in the hot water. After 5 minutes, read the temperature of the hot water and record it as the final temperature of the syringe contents. (The plunger tends to resist movement due to friction. Push in gently on the plunger before recording the final volume to make sure the plunger has not become stuck.) Record the final volume.

4. Repeat Step 3 using an ice-water bath.

5. How does compressibility differ among the three substances? Compare and contrast the compressibility data.

_____

_____

_____

_____

6. How does temperature-volume relationship differ among the three substances? Compare and contrast the temperature-volume data.

_____

_____

_____

_____

BedBugs

...Students observe the movement of BedBugs® under different conditions and relate this movement to the motion of particles in the solid, liquid, and gas states.

✔ Time Required

Setup	30	minutes the first time the activity is done
Performance	30	minutes
Cleanup	5	minutes

The BedBugs game

✔ Key Science Topics

- diffusion
- effusion
- molecular motion
- states of matter

✔ Student Background

Students should know that everything is made up of small particles and that these particles gain energy when a substance is heated and lose energy when a substance is cooled.

✔ National Science Education Standards

Science as Inquiry Standards:

- Abilities Necessary to Do Scientific Inquiry

  Students develop predictions, descriptions, explanations, and models using evidence.

  Students form a logical argument about the cause-and-effect relationship between energy applied and the behavior of the BedBugs, and relate this to the behavior of particles at the molecular level.

Physical Science Standards:

- Properties and Changes in Properties of Matter

  A characteristic property of matter is that as its temperature increases its particles increase their energy and move faster.

- Motions and Forces

  The motion of the BedBugs can be controlled by the kinetic energy.

- Transfer of Energy

  Energy is a property of all substances and is associated with the temperature and nature of a chemical substance. Energy is transferred and transformed in many ways, including heat energy being transformed to kinetic energy.

✔ Additional Process Skill

- making models   Students observe the behavior of the BedBugs as a model of the behavior of particles at the molecular level.

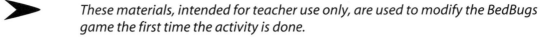

MATERIALS

For Getting Ready

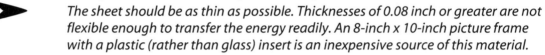

These materials, intended for teacher use only, are used to modify the BedBugs game the first time the activity is done.

- BedBugs game
- piece of heavy cardboard
- scissors
- glue
- BBs
- 2 size C batteries
- (optional) 2 pieces of coated copper wire
- (optional) 10-ohm rheostat
- (optional) thin, clear acrylic or Plexiglas™ sheet

The sheet should be as thin as possible. Thicknesses of 0.08 inch or greater are not flexible enough to transfer the energy readily. An 8-inch x 10-inch picture frame with a plastic (rather than glass) insert is an inexpensive source of this material.

- (optional) plastic-cutting knife, mat knife, or saw with plastic-cutting blade

Plastic-cutting knives are available at hardware stores.

For Introducing the Activity
Per class
- ice
- water
- clear container for heating water
- heat source

For the Procedure
Per class
- BedBugs game with or without rheostat prepared in Getting Ready
- simulated membrane pores/pinholes prepared in Getting Ready
- BB bugs prepared in Getting Ready
- (optional) piece of clear acrylic or Plexiglas cut to fit game "bed frame" prepared in Getting Ready
- (optional) overhead projector

For the Extension
Per class
All materials listed for the Procedure plus the following:
- double BedBugs with BBs prepared in Getting Ready

SAFETY AND DISPOSAL

For Getting Ready, follow all appropriate safety measures if using a plastic-cutting knife or other tools. No special safety or disposal procedures are required for the Procedure.

GETTING READY

1. (optional) You may wish to do the Procedure on the overhead projector so students can easily see what's happening. If so, cut the acrylic or Plexiglas sheet into a rectangle 14.0 cm x 21.5 cm (5½ inches x 8½ inches). Use the removable cardboard "blanket" from the BedBugs game's plastic "bed frame" as a guide. (Ignore the notches in the blanket.) Remove the cardboard blanket from the plastic frame and replace it with the piece of clear plastic.

2. (optional) Adding a rheostat to the circuit allows for finer and easier control of the amount of motion of the BedBugs. (See Figure 1.) Attach one end of each of the two coated copper wires to a different connector on the rheostat. (See ❶ in Figure 1.) On the BedBugs game, disconnect the red wire that runs from the motor to the battery at the battery end. (See ❷.) Replace that wire at the battery end with one of the pieces of coated copper wire from the rheostat. (See ❸.) Attach the other copper wire from the rheostat to the wire from the motor. (See ❹.) Be sure that the wires do not prevent the piece of cardboard or plastic from lying flat in the plastic frame.

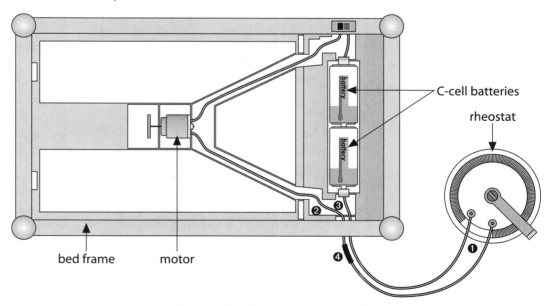

Figure 1: Attach the rheostat to the BedBugs game.

3. Make simulated membrane pores/pinholes for Parts C and D of the Procedure: Cut the piece of cardboard to be about 7½ inches x 2 inches. Mark off 5⅝ inches in the center of the cardboard as shown by the dotted lines in Figure 2. Fold the ends in along the dotted lines until

they are standing up perpendicular to the rest of the cardboard. The folded piece shown in Figure 3 should fit firmly between the edges of the plastic bed.

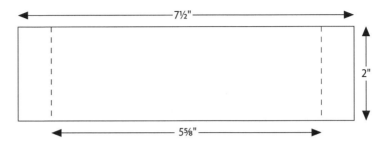

Figure 2: Cut the cardboard to size and fold on the dotted lines.

Cut notches in the bottom of the cardboard as shown in Figure 3. These notches represent "membrane pores" in Part C and "pinholes" in Part D. The holes should be approximately ¾ inch high. Their width is not critical. You may wish to make them narrow to begin with and enlarge them until you get a good rate of movement of the bugs through the holes.

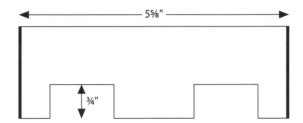

Figure 3: Cut notches in the cardboard divider.

4. For Part C, make several "BB bugs" by gluing two BBs inside each bug. Make all the BB bugs one color.

5. For the Extension, make four double BedBugs by gluing one bug on top of another and gluing BBs inside the bottom bug. Make all the double BedBugs one color.

INTRODUCING THE ACTIVITY

Provide samples of water in the solid, liquid, and gas phases. To generate the gas phase, heat water in a clear container so that the students can see that the amount of liquid water decreases. (Be sure that the students do not confuse the small drops of liquid water which condense as the water vapor cools above the container with the water vapor itself. Explain that water vapor is a colorless, invisible gas.) Lead a discussion of the similarities and differences between the three phases. Have the students suggest reasons for the differences. Tell them that you are going to show them a simulation which might help to clarify these differences.

PROCEDURE

Part A: Molecular Motion

1. (optional) If using the piece of clear plastic on the bottom of the bed, place the bed on the overhead projector.

2. Place a number of BedBugs in an ordered pattern on the game surface or piece of plastic. (See Figure 4.) Tell the students that the bugs represent particles of matter—that is, atoms or molecules.

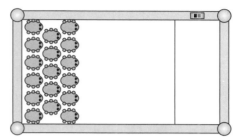

Figure 4: Place a number of bedbugs in an ordered pattern on the game surface.

3. Ask the students to predict what would happen to the bugs if you turned the game on and allowed the bed to vibrate slightly.

4. Before you turn the motor on, be sure that the vibration will be dampened (minimized) either by slightly lifting the game surface from below with your fingers or by using the rheostat. Turn on the motor and observe. (The bugs should quiver but not change position in their ordered pattern.) Turn the motor off and ask the students which state of matter the bugs represented (solid, liquid, or gas) and to explain their answer. *The bugs represented the solid state. Particles in the solid state are in a specific pattern. As energy is added, the particles vibrate more but they don't change position.*

5. Ask the students to predict what would happen to the bugs if you turned the game on and allowed the bed to vibrate a little more vigorously.

6. Turn the motor on and increase the vibration by slightly lowering the game surface or by turning the rheostat up slightly. Have the students observe. (The bugs will move more and the pattern will be lost.) Turn the motor off and ask the students which state of matter the bugs represented (solid, liquid, or gas) and to explain their answer. *The bugs represented the liquid state. The particles in the liquid state are less ordered than in the solid state. They move more and can change position because they have more energy.*

7. Ask the students to predict what would happen to the bugs if you turned the game on and allowed the bed to vibrate much more vigorously.

8. Turn the motor on and increase the vibration by releasing the game surface or by turning the rheostat up all the way. Have the students observe. (The bugs will move much more quickly and will bounce off the surface of the bed or, in some cases, totally out of the bed.) Turn the motor off and ask the students which state of matter the bugs represented (solid, liquid, or gas) and to explain their answer. *The bugs represented the gas state. Particles in the gas state move apart from each other and, as a result, they occupy more space. They have more energy than those in the solid and liquid states.*

9. Ask the students what increasing the vibration represents. *Heating. As you increase the temperature of an object, its kinetic energy increases and so does the movement of its particles.*

10. Ask the students how the simulation has confirmed or changed the reasoning they offered in Introducing the Activity for the differences between water in the three states.

Part B: Diffusion

1. Make sure that the game surface is flat and level.

2. Place bugs of one color at one end of the game surface and the same number of bugs of a second color on the other end. (See Figure 5.) Ask the students to predict the movement of the bugs when the motor is turned on. If they don't mention the distribution of the colors after the motor has been on for a minute or two, ask what sort of distribution they would expect.

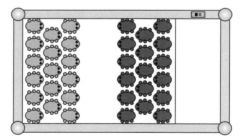

Figure 5: Place one color of bugs on one end and another color on the other end.

3. Turn the motor on and observe the movement of the bugs. (The bugs will move in a rapid random motion. They will mix, and each color should end up randomly distributed over the entire game surface. The effects of particle collisions are shown by the bugs bumping into one another.)

4. Tell the students that the name of the process being modeled here is diffusion. Diffusion is the mixing together of particles as a result of random thermal motion.

Investigating Solids, Liquids, and Gases with **TOYS**

Part C: Diffusion Through a Membrane

1. Show the students the cardboard partition. Explain that it represents a "porous membrane" such as the membrane that surrounds a cell or a balloon. The notches represent the pores in the membrane. Explain that diffusion can still take place through a porous membrane as particles move through the pores. Place the "porous membrane" between the plastic sideboards of the game surface, directly over the rotor on the motor. (See Figure 6.) It should fit snugly but not touch the game bed, so that the vibration will not be dampened. It should be high enough so that the bugs can pass through the openings. Make sure that the game surface is flat and level.

2. Place some regular bugs on one side of the "porous membrane" and BB bugs on the other side. (See Figure 6.) Ask the students to predict how the behavior with the motor on might be different from that observed in Part B.

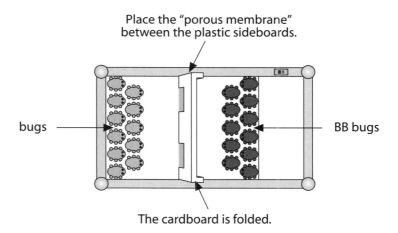

Figure 6: Place regular bugs on one side of the membrane and BB bugs on the other.

3. Turn the motor on and observe the movement of the bugs. (The regular bugs move at a faster rate and thus move through the openings more often than the BB bugs do. For example, in a given period of time, more regular bugs will pass through the holes than the number of BB bugs.)

4. Ask the students how the rate of diffusion through a membrane is affected by the mass of the particles. *The more massive the particles, the slower the rate of diffusion.*

Part D: Effusion

1. Explain to the students that the notches in the cardboard divider now represent pinholes, such as a tiny hole in a balloon. The bugs will only be able to get to the other side of the game board by going through the pinholes.

2. Place bugs of one color onto the game surface on one side of the "pinholes" (see Figure 7) and ask the students to predict what will happen to the bugs if the game is turned on.

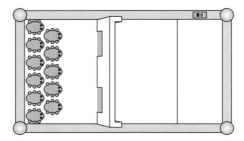

Figure 7: Set up the game surface for the effusion demonstration.

3. Turn the motor on and observe the movement of the bugs. *The bugs will move randomly, and some will pass through the pinholes.*

4. Tell the students that the name of the process being modeled here is effusion. Explain that effusion is related to, but not quite the same as diffusion. Diffusion refers to any spread of a substance throughout a space or throughout a second substance. Effusion specifically refers to the leakage of a gas through a tiny hole into a vacuum. If necessary to help students understand effusion, have the students consider the motion of a bee inside a car with the windows up. The bee will move in all directions, trying to get out. If a window is opened very slightly, the bee may escape.

● ● ● ● ● ● ● ● ● ● ● **EXTENSION**

Use the BedBugs to represent diffusion in a helium balloon. Place 12 single BedBugs of one color on one side of the divider to represent the helium in the balloon. Place four double BedBugs with BBs of another color on the other side of the divider to represent the nitrogen (N_2) and oxygen (O_2) in the air. (The double BedBugs represent diatomic molecules.) Turn on the motor and observe. Use the observations to explain why helium balloons get smaller over time. *A balloon is a porous membrane. The helium particles are less massive than the oxygen and nitrogen particles in air and thus the helium moves out of the balloon more rapidly than the oxygen and nitrogen move into the balloon. The balloon gets smaller with time because of the net loss of particles inside.*

EXPLANATION

➤ *The following explanation is intended for the teacher's information. Modify the explanation for students as required.*

Matter is made up of particles (atoms, molecules, and ions). The state a sample of matter is in depends on the properties of that matter and the amount of energy the sample has. When relatively little energy is present, the sample will be in the solid state. As its temperature is raised, more energy is added. If enough energy is added, the sample will melt. Still higher temperatures will further increase the energy of the particles, and the substance will eventually change to a gas. Figure 8 shows a schematic of the particles in each of the three phases.

solid   liquid   gas

Figure 8: This schematic diagram illustrates particles in the solid, liquid, and gas states.

In Part A of the Procedure, as the vibration of the BedBugs game increases, the amount of energy to the BedBugs increases and their movement increases. This is a visual analogy to the behavior occurring in the particles of the sample of matter described above as it changes from a solid to a liquid to a gas. (See Figure 9.)

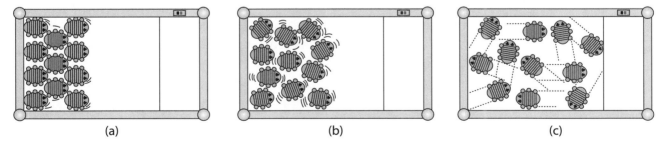

(a)   (b)   (c)

Figure 9: Bedbugs simulating (a) solid, (b) liquid, and (c) gas states.

The liquid and gas states of matter are fluids (materials that flow). In liquids and gases, the particles are free to move with respect to each other, which allows them to flow over and around each other. This ability to flow accounts for the fact that samples of liquids and gases take the shapes of their containers. Liquids have a definite volume because the particles that make them up are held in close proximity to each other due to attractive forces. Gases, however, have enough energy so that the particles are no longer attracted significantly by other particles but rather are spread out throughout the volume available to them in the container.

Parts B, C, and D of the Procedure model diffusion and effusion. The notched cardboard divider acts as the "porous membrane" or "pinhole" in Parts C and D.

Diffusion, the spontaneous mixing of the particles of two or more substances as a result of random thermal motion, can be shown with or without the divider. This mixing is clearly modeled in Part B, where bugs of one color start at one end of the game board and the same number of a second color start at the other end. After the motor has been on for a few minutes, the colors are mixed.

Using the divider in Part C demonstrates what happens when diffusion occurs through a semipermeable membrane. When both regular and BB bugs are used, the regular bugs move through the membrane more quickly than the ones with BBs. That is, the more massive the particles, the slower they move. This demonstration provides a qualitative visualization of Graham's law of diffusion. Simply stated, Graham's law shows that for a gas at a given temperature, the rate of diffusion is inversely proportional to the square root of the mass of the particles.

Effusion, the movement of gas particles through pinholes into a vacuum, is demonstrated in Part D. All the bugs start on one side of the divider. When the motor is turned on, the bugs gradually move through the holes in the divider. The bugs can pass through the divider in both directions, and the number of bugs on each side will gradually become constant even though movement is still occurring. The greater the mass of the gas particle, the slower the rate of effusion. If particles are mixed together, those with less mass will effuse at a faster rate. This qualitatively demonstrates Graham's law of effusion, which basically gives the same results as diffusion—the rate of effusion of particles is inversely proportional to the square root of their mass at a given temperature.

$$\frac{r_1}{r_2} = \sqrt{\frac{m_2}{m_1}}$$

ASSESSMENT

Options:

- (elementary) Ask the students to relate the states of matter to the ability to smell food cooking in the kitchen from another room. They can draw what happens and/or explain in words. *Cooking odors occur when some of the particles are heated sufficiently to go into the gas phase and diffuse through the air.*

- (middle school) Have students complete the Assessment Sheet (provided).

REFERENCE

Hogue, L.; Williams, J.P. "The BedBugs Game," *Journal of Chemical Education.* 1990, 67(7), 585–586.

HANDOUT MASTER

A master for the following handout is provided:
- Assessment Sheet

Copy as needed for classroom use.

BedBugs
Assessment Sheet

In each of the following scenarios, draw the BedBugs on the "After" diagrams as you would expect them to end up in each situation after the motor runs for a few minutes. Explain each of your drawings on another sheet of paper.

1. **Diffusion.** Each of these bugs has the same mass.

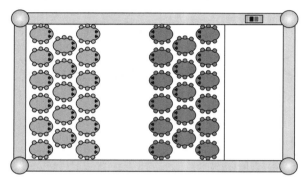 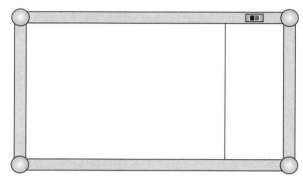

Before   After

2. **Diffusion through a membrane.** The bugs on the right are more massive because they are filled with BBs, and the divider is a porous membrane.

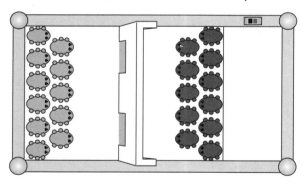 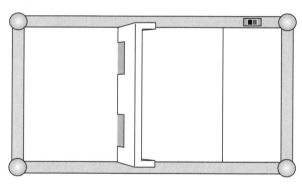

Before   After

3. **Effusion.** Each of these bugs has the same mass, and the holes in the divider are pinholes.

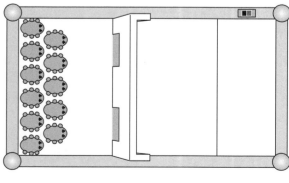 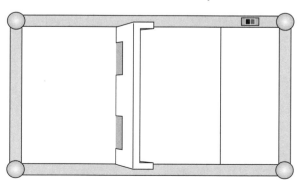

Before   After

Mystery Eggs

...Students investigate the properties of plastic eggs filled with solids, liquids, and gases and use these observations to hypothesize whether a chicken egg is hard-boiled or raw.

✔ Time Required

Setup	20	minutes the first time the activity is done
Performance	15–20	minutes
Cleanup	5	minutes

✔ Key Science Topics

• solid, liquid, and gas phases
• fluids

Some Mystery Eggs and their contents

✔ Student Background

Students should be familiar with the three common states of matter: solid, liquid, and gas.

✔ National Science Education Standards

Science as Inquiry Standards:

• Abilities Necessary to Do Scientific Inquiry

Students develop descriptions, explanations, and predictions based on what they observed about the behavior of eggs that contain different states of matter.

Students think critically about their observations of the eggs and form a logical argument about the cause-and-effect relationship in the investigation.

Physical Science Standards:

• Properties and Changes of Properties in Matter

The solid, liquid, and gas contained within the eggs each have properties that affect the behavior of the eggs when spun.

• Motions and Forces

The spinning motion of the eggs can be described by their position, direction, and speed.

✔ Additional Process Skills

• communicating	Students discuss their observations and interpretations of the behavior of the plastic eggs filled with solids, liquids, and gases.
• comparing/contrasting	Students compare the similarities and differences in the behavior of the plastic eggs filled with solids, liquids, and gases.
• investigating	Students manipulate the eggs and observe their behavior to predict if a solid, liquid, or gas is in each egg.

MATERIALS

For Getting Ready

Part A, per group

These materials are used to prepare the plastic eggs the first time the activity is done.

- 3 plastic eggs

The eggs must be the kind that can be opened and are empty inside. The three eggs should be different colors. You may wish to use one color for the "solid" eggs for every group, a second color for the "liquid" eggs for every group, and a third for the "gas" eggs. This facilitates discussion but is not necessary.

- 3 9- to 11-inch balloons

The best size depends on the size of the plastic eggs. The slightly inflated but not stretched balloon should just fit inside the egg.

Part A, per class
- sand
- (optional) funnel

Part B, per group
- 3 chicken eggs
- saucepan or other container to boil water
- stove or hot plate
- pushpin or needle
- container to catch blown-out egg contents
- liquid soap
- crayon or marker

For Introducing the Activity

Per class or group
- 9- to 11-inch balloon containing air
- 9- to 11-inch balloon containing water
- 9- to 11-inch balloon containing sand

For the Procedure

Part A, per group
- plastic egg filled with sand (See Getting Ready.)
- plastic egg filled with water (See Getting Ready.)
- plastic egg containing air-filled balloon (See Getting Ready.)

Part B, per group
- raw egg in shell
- hard-boiled egg in shell
- "blown" eggshell

For Variations and Extensions

❷ All materials listed for the Procedure plus the following:
Per group
- plastic eggs filled with different materials, such as the following:
  - rice or another solid with a density close to that of water
  - low-density material, such as Styrofoam™ or puffed rice

❸ All materials listed for the Procedure plus the following:
Per group
- marble, wooden, or hard-boiled egg

❹ Per group
- water
- corn syrup
- vegetable oil
- 3 plastic eggs containing 9- to 11-inch balloons filled with the 3 liquids listed for this extension

 The best size for the balloons depends on the size of the plastic eggs. The slightly inflated but not stretched balloon should just fit inside the egg.

❺ Per group
- 1 of the following toys:
  - snow scene or similar paperweight (also called a waterball)
  - compass which contains a liquid and the solid scale inside a clear sphere
  - "eye" or other slide ball (when the ball is rolled across the floor, the eye stays on the top)
  - clear sphere with a fish (or insect, etc.) floating inside
  - tornado tube

SAFETY AND DISPOSAL

Do not allow anyone to eat the eggs. Have students wash their hands after handling real eggs. Discard the hard-boiled and raw eggs. Keep the filled plastic eggs for future use.

GETTING READY

1. Prepare the set or sets of balloons for Introducing the Activity.

2. Fill one large balloon with sand for each group and pinch the neck of the balloon closed. (A funnel may help when filling the balloon.) The amount of sand needed will depend on the size of the plastic eggs— the sand-filled balloon should completely fill both halves of the plastic egg when the egg is closed. Place the sand-filled balloon into the larger (deeper) half of the plastic egg and continue holding onto the balloon neck. Check to see if you'll be able to close the egg by holding the other

half of the egg as tightly over the balloon as you can without letting go. If the balloon is too full, pour out some sand. Tie the balloon to seal it, and close the egg. (See Figure 1.)

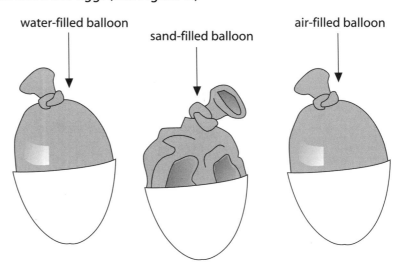

Figure 1: Fit the sand-filled, water-filled, and air-filled balloons into the plastic eggs.

3. Fill one water balloon with water for each group and pinch the neck of the balloon closed. (Do this at a faucet so that the water pressure forces the water into the balloon.) The water balloon should completely fill both sides of the plastic egg when the egg is closed. Check how well the water balloon fits into the egg as in Step 2, adjust if needed, tie the balloon, and close the egg.

4. Partially inflate one balloon with air for each group. Check how well the balloon fits into the egg as in Steps 2 and 3, adjust if needed, tie the balloon, and close the egg.

5. Hard-boil one egg for each group. Blow out another egg for each group as follows: Wash the egg thoroughly with soap and water. Punch a small hole in one end with, for example, a pushpin or needle, and a larger hole in the other end. Position the large hole over a container and blow through the small hole. (Be careful not to ingest any of the raw egg because of the risk of salmonella.) If nothing happens, push a pin or needle through the larger hole to break the egg membrane, and try again. The third egg should remain unaltered. Mark the three eggs in each set A, B, and C, using a crayon or marker.

INTRODUCING THE ACTIVITY

Give each group of students an air-filled balloon, a water-filled balloon, and a sand-filled balloon and ask them to record as many observations of each as possible (or prepare only one set of balloons and pass them around the class). Ask, "What state of matter do you think each balloon contains? What evidence can you cite to support your conclusions?" Lead a discussion of students' observations and point out that many of the differences between the balloons are a result of the different states of matter (solid, liquid, or gas) inside the balloons.

Discuss what techniques students used to figure out which state of matter was contained in each balloon. Most of the students will probably say that they squeezed the balloons. Discuss whether figuring out the states of matter in the balloons was easy or difficult. Tell students that they now have a more difficult challenge—to determine the states of matter inside rigid containers—in this case, plastic eggs.

PROCEDURE
Part A: Plastic Eggs

1. Show students a set of the three plastic eggs prepared in Steps 2–4 of Getting Ready. Explain that the goal of the investigation is to identify the state of matter contained within each egg (solid, liquid, or gas). To make this determination, the groups must investigate the eggs to determine the similarities and differences in their characteristics. Discuss the idea of a scientific investigation requiring systematic observation. Have students suggest ways that their observations of the eggs can be systematically made. Through discussion, be sure to bring out the idea that to fairly compare the eggs, the same observations must be made about each.

 In the following steps, if you wish to make the investigation more student-led, have students use a separate sheet of paper for recording their observations and make the Data Sheet optional.

2. Give each group a set of plastic eggs (an air-filled egg, a water-filled egg, and a sand-filled egg), but don't tell them which egg is which. Tell the students that they will have 10 minutes to conduct a scientific investigation of the eggs without opening them. Have students record their observations on the first section of the "Observing the Eggs" Data Sheet (provided).

3. Tell students that they will now make observations based on various tests of the eggs. Discuss the idea of controlling variables in these tests. Ask, "If we spin each egg on a different surface or apply varying amounts of force to spin each egg, can we conclude that differences in behavior are due to the contents?" Discuss how to avoid this problem. Some tests are

already listed in the table on the Data Sheet; have students come up with two or three additional tests of their own besides the ones that are listed. For their own tests, have students explain how they controlled variables. Have them record the results of all of their tests on the Data Sheet.

4. When the students have completed their initial investigations, lead a discussion of the groups' observations for each plastic egg. You may wish to record these on an overhead or on the chalkboard.

5. Explain to students that you now want them to try a "touch test." Have the students try spinning each egg, touching it just hard enough on top to bring it to a stop, and quickly removing their finger. (With some practice, they should observe that the solid egg stops completely and remains stopped, but the liquid egg will stop and then turn slightly again. In this case, the difference between the behavior of the eggs is subtle. Students may have to repeat it many times before they see the liquid egg move after stopping, and they may never see the gas egg move after stopping. See Figure 2.)

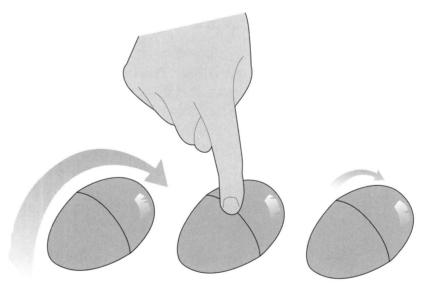

Figure 2: Touch the spinning liquid egg just hard enough to make it stop, and quickly remove your finger. The egg will turn slightly again.

6. Based on all of their observations and tests, have students record their proposed identifications of the phases of the eggs' contents and their evidence for each.

7. Have the students carefully open the plastic eggs to check their proposed identifications. Discuss the reasons for the eggs' behavior, introducing the following ideas. As much as possible, have students contribute their own ideas and observations as you conduct this discussion. Ask the students to describe any differences in the way the

plastic eggs spin. *When all three eggs are given equal amounts of spinning force, the solid egg spins fastest and longest. The liquid egg spins very slowly and stops spinning first. The gas egg wobbles the most while spinning and does not spin as long or as fast as the solid egg. The independent movement of the fluid contents causes the gas and liquid eggs to spin differently from the solid egg. When the liquid and gas eggs are stopped, the liquid or gas inside continues to move. When released, the plastic eggshell turns with the liquid. The impact of the gas is less because of the much smaller mass involved. Both of these phases are fluids because, unlike solids, they have the ability to flow.*

Part B: Real Eggs

1. Tell students that they are going to apply the techniques they learned in Part A to a new task: distinguishing between a raw egg, a hard-boiled egg, and a blown eggshell. Show students the raw egg in its shell, the hard-boiled egg in its shell, and the blown eggshell. Point out that one of these eggs contains (mostly) solid material, one contains (mostly) liquid material, and the third contains gas, and their job will be to determine the contents of each egg.

The blown egg is easy to recognize because of the hole in each end, but the others are more difficult to identify. Note that differences in mass will not help to distinguish between the hard-boiled and raw eggs, as the mass does not change significantly when the egg is boiled. Testing for fluid behavior by spinning the eggs works very well.

2. Have each group brainstorm how to distinguish between the raw and hard-boiled eggs. Have them test their ideas and form hypotheses about the eggs' contents. Although students will easily be able to identify the blown egg without testing, have them test this egg, too, just to see how it behaves under different conditions.

3. Once all students have predicted which of the chicken eggs' contents are blown out, hard-boiled, or raw, let them crack the eggs to check their predictions.

VARIATIONS AND EXTENSIONS

1. Set up the activity in a learning center and have students conduct the investigation using the Instruction and Observation Sheet (provided).

2. Students may assume that the heaviest egg always contains the solid. To counter this misconception, you may wish to introduce a fourth egg which contains a solid, such as rice, that has a density closer to that of the water. Have the students determine the state of matter in this egg. You may also wish to try eggs filled with a very low-density material such as Styrofoam or puffed rice. These eggs are difficult, if not impossible, to distinguish from eggs containing only air.

3. You may wish to provide younger students with a known (a solid marble or wooden "egg" or a hard-boiled egg) to aid them in identifying the plastic eggs' contents.

4. Tell the students that while all liquids are fluid, the rate of flow varies. Viscosity is a measure of the resistance to flow. The three liquids used in this extension have different viscosities. Ask the students to determine which egg contains the most-viscous liquid (corn syrup), which contains the least-viscous liquid (water), and which is in between (vegetable oil). They should explain the basis for their identification of the eggs. (The egg containing the most-viscous liquid will behave most like a solid-filled egg, while the egg containing the least-viscous liquid will behave most like a liquid-filled egg.)

5. Give each group one of the toys listed in Materials, have them observe its behavior, and hypothesize how the behavior relates to the main activity. The behavior of all of these toys is dependent on the flowing liquid inside. The snow scene paperweight and the tornado tube show very clearly that the liquid continues to flow after the solid container stops moving. The other toys allow the solid object to stay in or return to a particular position (eye on top, N towards the north, fish upright, etc.) because the solid floats in the liquid. These each also involve another scientific principle such as density or magnetism.

6. Have students do the "Which Egg Is Which?" Take-Home Challenge (handout provided) outside of school with an adult partner.

EXPLANATION

The following explanation is intended for the teacher's information. Modify the explanation for students as required.

Some of the properties explored in this activity are dependent only on the plastic egg containers and not on the contents; these include color, shape, and volume. Other properties are the result of the contents. For example, the egg full of sand has a greater mass than same-sized eggs with other contents, because sand is more dense than either water or air. If another solid is used, its density may or may not be greater than the density of the water. The gas-filled egg will always have lower mass and density than those filled with liquids or solids.

Spinning behavior is a useful way to distinguish between the less easily distinguishable solid and liquid eggs. When the plastic eggs are spun, the contents affect the observations. In the solid-filled eggs, the solid contents move together with the egg container or shell. The solid-filled eggs spin as a unit at a relatively fast rate and do not have the slowness and wobble seen with the liquid-filled egg and the wobbling of the gas-filled eggs.

When the solid egg is stopped, the sand stops with it, and the egg stays stopped. With the liquid- and gas-filled eggs, the fluid inside resists (but doesn't ignore) the spinning motion and moves somewhat independently of the container or shell. This accounts for the slower or more wobbly spin of the liquid- and gas-filled eggs. It also accounts for the observation that the liquid egg resumes spinning after the "touch test"; when the eggshell is stopped, the liquid inside remains in motion, and when the eggshell is released, the still-moving liquid pushes against the eggshell to start it moving again. This characteristic is most easily observed with liquid-filled eggs and is much more difficult to observe with gas-filled eggs because of the small mass of the latter.

The behavior of the filled plastic eggs correlates to the behavior of the real eggs. The contents of a raw egg are liquid, so they are able to move somewhat independently of the eggshell. Thus, the raw egg exhibits behavior similar to that observed with the water-filled plastic egg. When a real egg is hard-boiled, however, the contents become solid. Thus, the eggshell and contents move as a unit when the egg is spun, and the hard-boiled egg exhibits behavior similar to that observed with the sand-filled plastic egg.

HANDOUT MASTERS

Masters for the following handouts are provided:
- Observing the Eggs—Data Sheet
- Instruction and Observation Sheet
- Which Egg Is Which?—Take-Home Challenge

Copy as needed for classroom use.

Name _____   Date _____

Mystery Eggs
Observing the Eggs—Data Sheet

Record the color of each egg at the top of one of the columns and your observations of the properties of each colored plastic egg below. (The blank rows are for tests your group comes up with.) When all tests are complete, record your proposed identity of the state of matter inside each egg. Explain your conclusions on the back of the sheet, using your observations as evidence. Open the eggs and record the actual state of matter inside each.

	Egg 1: _____ (color)	Egg 2: _____ (color)	Egg 3: _____ (color)
Initial Observations			
Sound when shaken			
Feel when shaken			
Relative mass			
Tests			
What happens when you roll the egg?			
What happens when you spin the egg?			
What happens when you stand the egg on one end?			
Touch test			
Predicted state of matter inside egg			
Actual state of matter inside egg			

Reproduced from *Investigating Solids, Liquids, and Gases with TOYS*, published by Terrific Science Press.

Mystery Eggs
Instruction and Observation Sheet

1. Your goal in this investigation is to identify the state of matter contained within each egg. Conduct systematic observations of the set of plastic eggs for about 10 minutes. Be sure to compare the eggs consistently; that is, evaluate the same set of characteristics for each egg. Record your observations on the Data Sheet.

2. Now you'll need to conduct tests of the eggs to make further observations. Be sure to control the variables in your tests. For example, if you spin the eggs on different surfaces or apply varying amounts of force to spin the eggs, you cannot conclude that differences in behavior are due to the contents of the eggs. Conduct the tests listed on the Data Sheet and devise two or three additional tests. Record the results of all your tests on the Data Sheet. For the "touch test," spin each egg on its side, touch each just hard enough on top to bring it to a stop, then quickly remove your finger. Try this several times for each egg and observe carefully. The differences may be very slight.

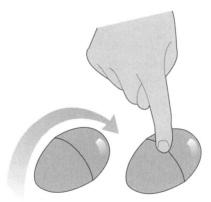

Touch the spinning egg just hard enough to make it stop, and quickly remove your finger.

For your own tests, explain how you controlled the variables. _____

_____

_____

_____

3. Based on all your observations and tests, predict the state of matter in each egg and record your predictions on the Data Sheet.

Discuss the evidence you have to support your predictions. _____

_____

_____

_____

Mystery Eggs

Which Egg Is Which?—Take-Home Challenge

Date: _____

Dear Adult Partner(s):

Our class has been learning about the behavior of solids, liquids, and gases in egg-shaped containers. We would like you to participate in this learning experience by carrying out the challenge of distinguishing between a raw egg and a hard-boiled egg without cracking them. What we as a class are trying to determine is how different people would approach this problem—for example, what experiments would they try, and what types of reasoning would they use? Therefore, your child will not describe our investigation until you've finished the take-home challenge. The students will add the results of this take-home challenge to our class results to provide us with additional data.

In Advance
Gather two chicken eggs of about the same size. Hard-boil one. (If the egg cracks while being boiled, start over with another egg.) Let the hard-boiled egg cool and place both eggs in the refrigerator so they are the same temperature. Have your child get the eggs out of the refrigerator and mix them up so that you do not know which one is which.

The Challenge
Your challenge is to determine which egg has been hard-boiled and which is still raw without breaking the eggshells. Your child will act as recorder of your experiments and briefly note the results in the space below or on the back of this page. After you think you know which is which, explain your reasoning to your child, who should record a brief summary of your explanation.

A Solution
Have your child show you what he or she learned in class about the behavior of the eggs and explain to you why the different eggs behave as they do.

Thank you for helping us with our class research project. Have fun!

Sincerely,

_____

Balloon in a Bottle

…Students discover that air takes up space.

A balloon in a bottle

✔ Time Required

Setup	5	minutes
Performance	10	minutes
Cleanup		none

✔ Key Science Topics

- nature of gases
- properties of air

✔ Student Background

This activity is a good attention getter when discussing air or gases.

✔ National Science Education Standards

Science as Inquiry Standards:

- Abilities Necessary to Do Scientific Inquiry

  Students conduct a simple investigation and use the evidence from that investigation to propose explanations.

Physical Science Standards:

- Properties and Changes of Properties in Matter

  A characteristic property of air is that, although invisible, it takes up space.

✔ Additional Process Skills

• inferring	Students infer that air takes up space because it prevents inflation of a balloon in a bottle.
• hypothesizing	Students hypothesize methods of inflating the balloon.

MATERIALS

For Getting Ready

The sharp object, intended for teacher use only, is needed to make holes in some of the soft-drink bottles the first time the activity is done.

- pushpin, thumbtack, or nail

For Introducing the Activity
- balloon
- paper bag, beach ball, or inner tube

For the Procedure

Per group
- balloon
- clear plastic 1-L soft-drink bottle

Although 1-L bottles work best, 2-L bottles can be substituted.

- (optional) extra balloons

For Variations and Extensions

❷ All materials listed for the Procedure plus the following:
Per class
- soldering iron, hot-melt glue gun, or hot nail with a hot pad to make hole
- rubber stopper to fit the hole (See Figure 2.)
- bucket

❹ All materials listed for the Procedure plus the following:
Per class
- scissors or knife

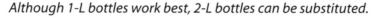

SAFETY AND DISPOSAL

For health reasons, each student blowing up a balloon must have his or her own previously unused balloon. No special disposal procedures are required.

GETTING READY

1. Push a deflated balloon into each of the bottles and stretch the open end of the balloon back over the bottle's mouth to seal the bottle. (See Figure 1.)

2. For Part B, make holes in some of the plastic bottles by carefully pushing a pushpin, thumbtack, or nail through the side.

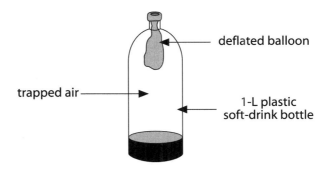

trapped air

deflated balloon

1-L plastic
soft-drink bottle

Figure 1: Prepare the balloon in a bottle.

INTRODUCING THE ACTIVITY

The activity could be introduced by blowing up a balloon in front of the class and discussing how the balloon expands as air is pushed into it during the blowing process. Blowing up a balloon is an example of air taking up space. Tie off the balloon and gently squeeze it to show that the balloon cannot be completely flattened without popping it. Challenge the students to explain why—but don't tell them the answer yet. You may want to pop the balloon to show it can indeed be flattened. To further illustrate the fact that air takes up space, blow up other items (such as a paper bag, beach ball, or inner tube).

PROCEDURE

Part A: The Challenge

1. Divide the class into groups. Let each group select a representative who will blow up the balloon.

   Make sure the students understand that only one student should blow into a balloon (to avoid spreading germs).

2. Ask the groups to predict what will happen when their representative tries to blow up the balloon in the bottle. Groups should record their predictions on the Instruction and Observation Sheet (provided).

3. Distribute a balloon and bottle (without a hole) setup to each representative.

4. Challenge each group representative to blow up the balloon inside the bottle, and have him or her describe to the group the resistance felt when trying to blow up the balloon. Other members of the group should describe what they saw.

   No matter how hard the students try, the balloons will not inflate very much.

5. Have each representative remove the balloon from the bottle and blow it up to show that the balloons can be inflated.

6. Have the groups propose explanations for what happened, supporting their explanations with observations.

7. (optional) Give a new, previously unused balloon to each member of each group. Allow each person to try inflating the balloon in the bottle. After each trial encourage the groups to reconsider their explanations to allow for refinement based on multiple experiments.

Part B: The Hole

1. Using the bottles with holes that were prepared in Getting Ready, repeat Part A and challenge students to explain the difference between the observations in these two parts. *The balloon will inflate in the bottle with a hole because the air is no longer trapped inside the bottle.*

2. Ask, "How can we easily test that idea?" Students will probably suggest putting a finger over the hole before blowing. Have each representative try this idea. Ask students to describe what happened. *Once again, the trapped air prevents the balloon from being blown up very much.* Ask, "Does this support your explanation? Why or why not?"

3. Have each representative blow up the balloon with the hole open. (The balloon inflates.) Have each representative keep his or her mouth over the balloon opening. Will the balloon stay inflated? *Yes, as long as the mouth is sealing in the air.* Have each representative place a finger over the hole and remove his or her mouth but keep the finger over the hole. What happens? *The balloon stays inflated.* Ask, "Based on what you have observed previously, how would you explain why the balloon stays inflated? How could you test your idea?" Allow students to propose explanations and test their ideas.

VARIATIONS AND EXTENSIONS

1. Set up the activity in a learning center and have students conduct the investigation using the Instruction and Observation Sheet (provided).

2. Make a balloon water-shooter by following this procedure:

   a. Make a hole approximately 1 cm in diameter in the side of a 1- or 2-L bottle just above the base. (See Figure 2a.)

   > Use a hot nail, a soldering iron without solder, or a hot-melt glue gun without glue to melt a smooth hole in the bottle which can be sealed with a rubber stopper.

   b. Put a balloon in the bottle and stretch the open end of the balloon back over the bottle's mouth.

c. Blow up the balloon. (See Figure 2b.) With your mouth still in place and the balloon still inflated, plug the hole with the stopper. (See Figure 2c.) Remove your mouth and the balloon will stay inflated. (See the photo on the first page of this activity.)

The bottle may collapse slightly around the balloon when you remove your mouth, or the balloon may shrink slightly. This will not affect the results.

d. Fill the balloon with water. Go outside (where water spills will not matter) and set up a bucket as a target.

e. Aim the bottle at the target and take the stopper out. Water will be pushed out of the balloon.

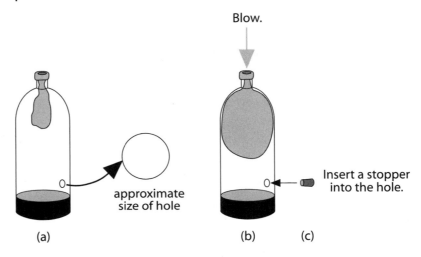

Figure 2: Make a balloon water-shooter.

3. Put a balloon inside a plastic 2-L bottle without a hole as in Figure 1. Squeeze the bottle firmly. The balloon will be pushed out of the bottle and become inflated and may even slip off the bottle.

4. Construct a model of a working lung: Cut off the bottom half of a 1-L bottle. Tie an uninflated balloon, then cut off the other end. (See Figure 3a.) Stretch the cut balloon across the cut end of the 1-L bottle; tension should hold it in place. (See Figure 3b. This balloon represents the diaphragm.) Place an intact uninflated balloon inside the top of the 1-L bottle and stretch the open end of the balloon back over the mouth to seal the bottle. (See Figure 3c. This balloon represents the lung.) Pull on the balloon that is over the cut-off end of the bottle. As the pressure in the bottle drops, air will be pushed into the intact balloon, simulating the contraction of the diaphragm and the corresponding expansion of the chest cavity and lungs. (See Figure 3d.) As you release the cut-off balloon, some of the air is expelled from the "lung" balloon, similar to when the diaphragm relaxes and air is forced out of the lungs. While the volume change in this demonstration is not nearly as great as in a real lung, it should be easily visible.

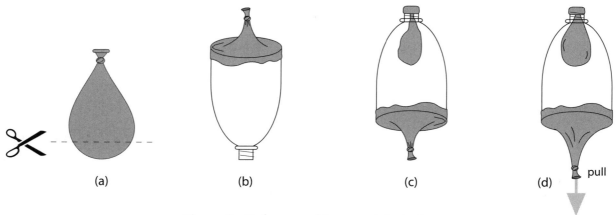

(a)   (b)   (c)   (d) pull

Figure 3: Make a working model of a lung.

EXPLANATION

The following explanation is intended for the teacher's information. Modify the explanation for students as required.

Blowing up a balloon involves forcing additional air into the balloon. The gas particles blown into the balloon hit the inside walls of the balloon, creating enough pressure to force the rubber of the balloon to expand and the balloon to inflate. The collision of these particles with the walls creates the pressure inside the balloon. There is also pressure outside the balloon (atmospheric pressure). Atmospheric pressure is a result of molecules of gas in the atmosphere pushing on an object. For the balloon to stay inflated, the pressure inside the balloon must be greater than atmospheric pressure. Additional pressure is necessary because the pressure inside the balloon must not only counter the atmospheric pressure but also stretch the elastic rubber when a balloon is inflated. If the mouth of the inflated balloon is opened (or the balloon pops), the extra air inside will quickly flow out because gases move from areas of higher pressure to areas of lower pressure.

Part A of the activity shows that it is impossible to significantly inflate the balloon by mouth when the balloon is inside a closed bottle. (See Figure 4.) The pressure of the air trapped inside the bottle prevents you from inflating the balloon.

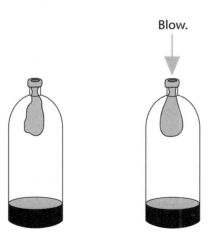

Blow.

Figure 4: It is impossible to significantly inflate a balloon by blowing into it when it is inside a closed bottle.

In order to actually blow up the balloon in the bottle, you not only need to blow enough air into the balloon to provide the pressure needed to stretch the rubber of the balloon, but you also have to apply enough pressure to compress the air trapped in the bottle. This compression is needed to make room for the inflated balloon. Even though gases are compressible, it is difficult for most of us to exert enough pressure just by blowing to compress the trapped air very much. (See Figure 5a.)

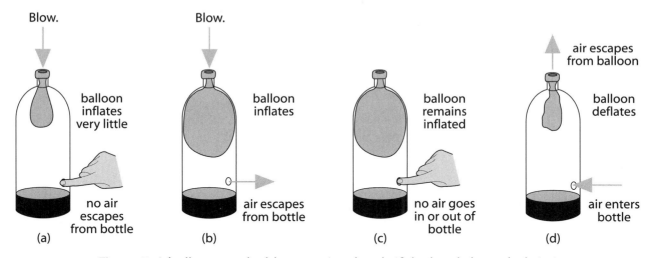

Figure 5: A balloon can be blown up in a bottle if the bottle has a hole in it.

A balloon can be blown up in a bottle if the bottle has an opening to allow the air initially inside the bottle to escape to make room for the expanding balloon. (See Figure 5b.) If you seal the hole with your finger or a stopper after inflating the balloon, the balloon will remain inflated even after you remove your mouth. (See Figure 5c.) The careful observer may notice that the bottom part of the plastic bottle contracts slightly around the inflated balloon or the balloon shrinks very slightly. The atmospheric pressure exerted on the open end of the balloon is greater than the pressure in the region between the balloon and the bottle. This larger pressure keeps the balloon inflated.

The balloon will deflate to its original size when the hole in the side of the bottle is opened again. (See Figure 5d.) This is because opening the side hole again allows the air pressure on both sides of the balloon to equalize.

ASSESSMENT

Options:

- Have students demonstrate the experiment to a younger class so that these younger students are able to show the experiment to their parents and family.

- Have students brainstorm new and different experiments they could do to demonstrate what they learned about air in this activity. Have the students present their demonstrations to the class.

CROSS-CURRICULAR INTEGRATION

Life science:
- Have students research how the lungs work and compare this to the action of a balloon in a bottle.

REFERENCES

"The Bottled Balloon," *SuperScience.* October 1991, 28–29.
Ontario Science Center. *Scienceworks;* Addison-Wesley: Reading, MA, 1984; p 6.

CONTRIBUTORS

Mark Beck, Indian Meadows Primary School, Ft. Wayne, IN; Teaching Science with TOYS peer mentor.
Carol Schelbert, Randall School, Peru, IN; Teaching Science with TOYS, 1993–94.

HANDOUT MASTER

A master for the following handout is provided:
- Instruction and Observation Sheet
Copy as needed for classroom use.

Balloon in a Bottle
Instruction and Observation Sheet

PART A

1. Push a deflated balloon into the bottle without a hole and stretch the open end of the balloon back over the mouth of the bottle. (See figure.) Be sure you use your OWN balloon. Predict what will happen when you blow into the balloon. _____

    _____

2. Now blow into your balloon. What happened?_____

    _____

3. Take your balloon out of the bottle and blow into it. What happened? _____

    _____

4. Can you explain the difference between what happened in Step 2 and Step 3? Discuss the result with your group. _____

    _____

    _____

PART B

5. Put your balloon in the bottle with the hole. Predict what will happen this time when you blow into the balloon. _____

    _____

6. Now blow into your balloon. What happened? Can you explain this? _____

    _____

    _____

7. What did you learn from this activity? _____

    _____

    _____

Burping Bottle

...Students conclude that air takes up space.

✔ Time Required

Setup   5 minutes
Performance   10 minutes + time to
                  investigate in Part B
Cleanup   5 minutes

✔ Key Science Topics

- density
- gases
- gravity
- liquids

Burping Bottle

✔ National Science Education Standards

Science as Inquiry Standards:

- Abilities Necessary to Do Scientific Inquiry

  Students identify questions about the Burping Bottle that can be answered through scientific investigation.

  Students design and conduct a scientific investigation using variations of the basic Burping Bottle procedure and/or apparatus.

  Students analyze and interpret the data they collect.

Physical Science Standards:

- Properties and Changes of Properties in Matter

  A characteristic property of air is that, although invisible, it takes up space.

✔ Additional Process Skill

- observing   Students watch and listen during the activity.

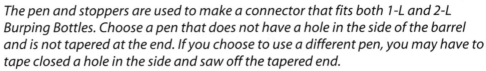

MATERIALS

For Getting Ready
Per group as a hands-on activity or per class as a demonstration
- barrel of a pen
- 2 #3 1-hole rubber stoppers

The pen and stoppers are used to make a connector that fits both 1-L and 2-L Burping Bottles. Choose a pen that does not have a hole in the side of the barrel and is not tapered at the end. If you choose to use a different pen, you may have to tape closed a hole in the side and saw off the tapered end.

- (optional) soapy water

For Introducing the Activity
- rigid plastic or glass bottle
- sink or bucket

For the Procedure
Part A, per group as a hands-on activity or per class as a demonstration
- funnel made from the top of a plastic 2-L soft-drink bottle
- plastic 2-L soft-drink bottle
- connecting device made in Getting Ready
- container of water
- shallow tray
- (optional) food color

Part B, per group as a hands-on activity or per class as a demonstration
- simple variations of basic Burping Bottle

SAFETY AND DISPOSAL

No special safety or disposal procedures are required.

GETTING READY

1. Make a connecting device for the Burping Bottle. Push one end of the pen barrel through the hole in the rubber stopper from top to bottom. Make sure the barrel of the pen will slide completely through the hole in the stopper. If it does not, enlarge the hole. Wetting the hole with soapy water can make it easier to insert the barrel of the pen. Push the other end of the pen barrel completely through the hole in the other stopper from top to bottom. Both ends of the pen barrel should protrude about 1 cm from the smaller ends of the stoppers. (See photo on the first page of this activity and Figure 2.)

As an alternative, use a hot-melt glue gun without glue to melt a hole in each of two 2-L bottle caps. The hole must be large enough for a straw to slide through.

Insert a straw through the holes and use hot-melt glue to secure the tops of the bottle caps together. Place a cut-off piece of a 35-mm film container over the bottle caps to reinforce the connector.

2. If you are doing the activity as a demonstration, assemble the apparatus in advance by following the steps listed in the Procedure.

INTRODUCING THE ACTIVITY

Show students how water runs out of a rigid bottle when held upside down. (It comes out in spurts.) Ask students to describe what happened. Then show them that the water drains without spurting (and more rapidly) if it is swirled to create a vortex that allows air to continuously enter the bottle while the water is draining. (See Figure 1.) Ask students to describe what happened this time. Have students suggest reasons for the difference between the behavior of the water in the two situations. Accept student ideas. Tell them that you will be discussing their ideas further after they observe the Burping Bottle.

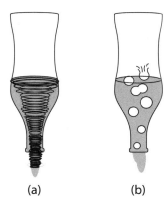

(a)   (b)

Figure 1: Water can drain from a bottle more quickly (a) if it is swirled into a vortex than (b) if it is just held upside down.

PROCEDURE
Part A: Make a Bottle Burp

1. Fill the soft-drink bottle half-full of water. Ask students, "Is the bottle full?" They will probably say that it is half-full, referring only to the water. Lead them to realize that the bottle is actually full, half with water and half with air.

2. Attach the connecting device to the bottle and funnel, and secure it tightly. (See Figure 2.)

3. (optional) Add food color to the container of water and stir.

4. Place the apparatus on a shallow tray.

5. Pour water into the funnel until it is two-thirds full.

6. Wait a minute. If the "burping" action does not begin, squeeze the soft-drink bottle slightly until an air bubble escapes from the pen barrel into the funnel. Your Burping Bottle is now in operation. Ask students to describe what happened. Ask, "Where did the bubble come from?" *The bottle on the bottom.* "What happens when air moves from the bottle to the funnel?" *Water flows into the bottle.*

7. Keep adding water to the funnel until you wish the operation to stop (or the bottle is full of water).

8. Remind students of the inverted bottle of water in Introducing the Activity. Discuss how the flow of water from an inverted bottle relates to the Burping Bottle. Ask, "What can you infer about air in both cases?" *Air takes up space.*

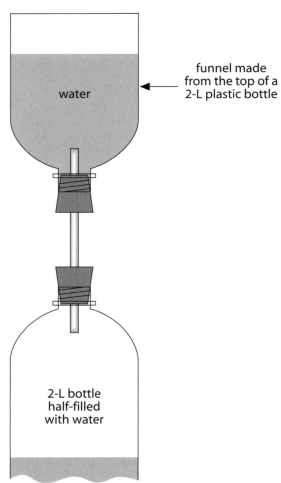

Figure 2: Assemble the Burping Bottle.

Part B: Investigating the Variables

1. Ask students, "Now that you've seen it in action, what variation could be made that would affect the burping action?" Record all ideas on a chalkboard, overhead, or chart.

2. As a class, come up with a list of variables you or students could test with the existing apparatus or simple variations of it and a list of outcomes that one or more of the variables could possibly affect. For example, students could vary the size of the bottle, the shape of the bottle, the volume of water in the bottle, the proportion of the bottle filled, and/or the size of the funnel. Outcomes that may be affected by these variables include the interval before the first burp, the interval between burps, and/or the time the funnel takes to empty or the bottle to fill.

3. Have students conduct their investigations, make observations, and share their outcomes.

VARIATION

Set up the activity in a learning center and have students conduct the investigation using the "Observing the Bottle" Instruction and Observation Sheet (provided).

EXPLANATION

> *The following explanation is intended for the teacher's information. Modify the explanation for students as required.*

The soft-drink bottle in this activity is filled with air and water. Air and water are both matter and take up space. Since water is more dense, it is at the bottom of the bottle. Even though the water in the funnel is more dense than air, it cannot enter the soft-drink bottle until some air is released. Notice that no water enters the bottle until a bubble of air rises in the tube. The burping process will continue as long as there is water in the funnel or until the bottle is full of water.

If no water is present in the bottom bottle to begin with, the first burp occurs very quickly after pouring begins. If a bottle smaller than 2 L is used, burps will typically occur at shorter intervals and will be smaller in volume than burps produced in a 2-L bottle.

ASSESSMENT
Options:

- Ask students to write a paragraph describing how the Burping Bottle works.

- Evaluate students' completed Instruction and Observation Sheets, paying particular attention to their understanding of the difference between observations and explanations.

CROSS-CURRICULAR INTEGRATION

Earth science:
- Have students relate this activity to the mud pots in Yellowstone National Park.

Life science:
- Have students compare the burping in this activity to burping in humans.

REFERENCE

Sarquis, A.M.; et al. "Take-Home Challenges: Extending Discovery-Based Activities Beyond the General Chemistry Classroom," *Journal of Chemical Education.* 1996, *73* (4), 337–338.

CONTRIBUTORS

Ed Sacha, McGregor Elementary School, Detroit, MI; Institute for Chemical Education participant.

Pam Mason, Research Associate, Center for Chemical Education, Miami University, 1995.

R. Dean Bernard, Wilmington High School, Wilmington, OH; Teaching Science with TOYS, 1992–93.

JoAnne Lewis, Stanberry Elementary School, Stanberry, MO; Teaching Science with TOYS, 1994.

Jane Newcomer, Stanberry Elementary School, Stanberry, MO; Teaching Science with TOYS, 1994.

Paul Schumm, Ayersville High School, Defiance, OH; Teaching Science with TOYS, 1995.

HANDOUT MASTER

A master for the following handout is provided:
- Observing the Bottle—Instruction and Observation Sheet

Copy as needed for classroom use.

Burping Bottle

Observing the Bottle—Instruction and Observation Sheet

PART A

1. Fill a soft-drink bottle half-full of water. Attach the connecting device to the bottle and funnel, and secure tightly. If you wish, add food color to the container of water and stir. Place the apparatus on a shallow tray. Pour water into the funnel until it is two-thirds full. Wait a minute. If the "burping" action does not begin, squeeze the soft-drink bottle slightly until an air bubble escapes from the pen barrel into the funnel. Your Burping Bottle should now be operational.

2. Describe the action of the Burping Bottle. In your discussion, answer questions such as, "Where did the bubble come from?" and "What happens when air moves from the bottle to the funnel?"

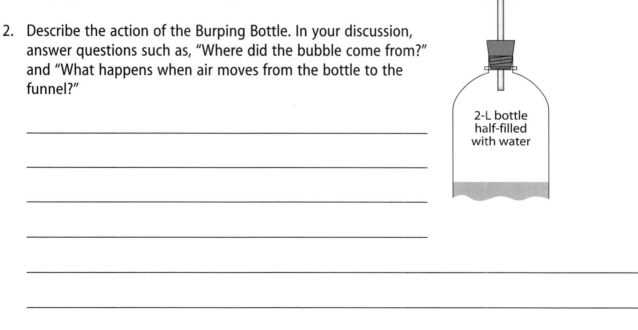

water

funnel made
from the top of a
2-L plastic bottle

2-L bottle
half-filled
with water

   _____

   _____

   _____

   _____

   _____

   _____

   _____

3. Keep adding water to the funnel until you wish the operation to stop (or the bottle is full of water).

PART B

Using the basic Burping Bottle procedure as a starting point, design your own investigation. First, decide what question you want to answer. The question must be one you can answer by making observations and collecting data using the Burping Bottle. For example, you might want to know how the amount of water you start with in the bottle affects the burping behavior. Describe your investigation and its outcome below.

1. What question do you want to answer?

   _____

2. How will you set up your investigation to answer this question?

   _____

   _____

   _____

3. What observations will you make and what data will you collect?

   _____

   _____

   _____

4. What happened?

   _____

   _____

   _____

5. Use your observations and data to explain why you saw the result you did.

   _____

   _____

   _____

   _____

   _____

Tissue in a Cup

…How can you keep a tissue dry under water?

✔ **_Time Required_**

Setup   5   minutes
Performance   10–15 minutes*
Cleanup   5   minutes
*Additional time is required if done as a hands-on activity.

✔ **_Key Science Topics_**

- air takes up space
- matter

✔ **_National Science Education Standards_**

Science as Inquiry Standards:

- Abilities Necessary to Do Scientific Inquiry
  Students use observations as evidence for explanations.

Physical Science Standards:

- Properties and Changes of Properties in Matter
  A characteristic property of air is that, although invisible, it takes up space.

✔ **_Additional Process Skill_**

- comparing/contrasting   Students compare and contrast the results of pushing the cup under the water with the hole closed and opened.

MATERIALS

For Getting Ready
Per class
- 1 of the following:
  - pushpin
  - nail, heat source, and pliers
  - hot-melt glue gun

For the Procedure
Per group or class
- clear, 10-ounce, soft plastic cup with hole prepared in Getting Ready

➤ *Solo® makes a cup like this. The letters PETE and recycling code 1 are on the bottom of the cup. Don't use rigid plastic cups; they will crack when you try to make the holes.*

- 4 or 5 facial tissues
- food color
- (optional) tape
- 1 of the following large-mouthed containers (preferably clear):
  - small, plastic aquarium
  - plastic bucket
  - large beaker

➤ *You can make a large beaker by cutting the top quarter off of a plastic 2- or 3-L soft-drink bottle.*

For Variations and Extensions
❶❷ All materials listed for the Procedure plus the following:
Per class
- (optional) oil-based clay

❸ All materials listed for the Procedure plus the following:
Per class
- containers of various shapes, such as wide-mouthed cups, bowls, plastic bags
- containers made of different materials, such as Styrofoam™, plastic, and cardboard

❹ All material listed for the Procedure plus the following:
Per class
- additional facial tissues

❺ All materials listed for the Procedure except:
Per class
- substitute the cup with hole prepared in Getting Ready with cups designed by students

Investigating Solids, Liquids, and Gases with **TOYS**

SAFETY AND DISPOSAL

No special disposal procedures are required for the Procedure. If using a nail to make the holes in the cups in Getting Ready, use caution when heating the nail and when handling it.

GETTING READY

If using a pushpin, push the pushpin with a twisting motion through the bottom of an empty cup until the cup is punctured. (See Figure 1.) If using a nail, hold the nail with insulated pliers while carefully heating it. When the nail is hot, push it through the bottom of an empty cup using the pliers. Remove the pushpin or nail. If using a hot-melt glue gun, use the tip of the gun (without glue) to melt a small hole in the bottom of the cup. (The size of the hole will affect how quickly you see results in Step 7 of the Procedure.)

Figure 1: Push a pushpin or nail through the bottom of a cup.

INTRODUCING THE ACTIVITY

Ask students to suggest reasons why keeping something dry under water might be important. Invite students to suggest methods for doing so. Tell students you have a way to keep something dry that depends on air taking up space.

PROCEDURE

1. Crush tissues and fit them tightly into the bottom of the plastic cup so that they won't fall out when the cup is turned upside down.

   You may need to use tape to hold the tissue in place, but be sure the tape does not cover the hole in the bottom of the cup.

2. Ask students to predict what will happen when the inverted cup containing the tissue is submerged in water.

3. Put your finger firmly over the hole in the bottom of the cup. With the cup upside down, push the cup down into a container of colored water until it is completely submerged. (See Figure 2.)

➤ *You must keep the cup very straight as you push.*

4. Keeping your finger on the hole, withdraw the plastic cup from the water.

➤ *Remember to keep the cup straight as you lift it.*

5. Dry the lip and outside of the cup with a paper towel and remove the tissue from the cup.

6. Let students feel the tissue and report their observations.

7. Leaving your finger off the hole on the bottom of the cup, repeat Steps 1–6. Let students feel the tissue and report their observations.

➤ *If the hole in the cup is small, you may need to submerge it a while before something happens.*

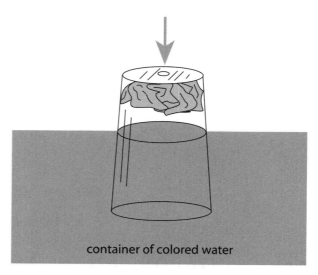

container of colored water

Figure 2: Push the cup under water.

8. Ask students to propose explanations for why the tissue stayed dry, using their observations as evidence. Discuss student ideas, and provide additional information and clarification if needed.

VARIATIONS AND EXTENSIONS

1. Set up the activity in a learning center and have students conduct the investigation using the Instruction and Observation Sheet (provided). Young students may need to use a clay ball to cover the hole.

2. Try using a cup with three holes punched in the bottom. Use tape or clay to close the holes. When the cup is submerged, open all three holes. How quickly does the water fill the cup? Compare this with the cup that has only one hole. Students can predict which tissue will get wet first. Students can also estimate how many seconds it takes to get the tissue wet.

3. Try this activity using different-shaped containers (wide-mouthed cup, bowl, plastic bag). Try containers made of different materials (Styrofoam, plastic, cardboard). Do the results change? Why or why not?

4. Ask students to estimate and then test the number of tissues that can be put in the cup and successfully kept dry in this activity. (Pack them in tight!) Have the class create a histogram of the maximum number of tissues each group could successfully keep dry.

5. Challenge students to design a cup that will fill halfway with water, but no more, when submerged upside down. (Place the hole in the side of the cup, halfway up. Water will enter the bottom of the cup and force air out the hole in the side of the cup. When the water level reaches the hole, no more air will escape, so no more water can enter and the cup will be filled halfway.)

EXPLANATION

The following explanation is intended for the teacher's information. Modify the explanation for students as required.

Air is matter, and even though we can't see it, it takes up space. Air is trapped in the cup when you submerge it with your finger over the hole in its bottom. The air in the cup keeps water out, so the tissues stay dry (unless the tissues slip downward). Some water will enter the open end of the cup because the air in the cup is slightly compressed as the cup is pushed into the water.

Repeating the procedure with the hole open allows the air to be pushed out of the cup through the hole as water enters the cup. Since air can escape, water fills the cup, and the tissues are wetted.

ASSESSMENT

Tell students that divers used to use diving bells to enable them to breathe underwater. Diving bells were large, hollow, open-ended structures that were lowered into the water, open end down. Ask students to use words and illustrations to explain why divers could breathe using diving bells. (For more information about diving bells, see the social studies section in Cross-Curricular Integration.)

CROSS-CURRICULAR INTEGRATION

Social studies:

- Ask students to describe the types of activities that people might do underwater that would require them to take their own air with them. *Scuba diver, person in a submarine.*

- Discuss the historic use of the diving bell, which is a good example of how air takes up space. A hollow structure at one end is lowered into water with the opening pointing downward, trapping a pocket of air that can be used for breathing by passengers. (See Figure 3.) The earliest record of a diving bell appeared in the 1500s, but the forerunner of the modern diving bell was designed by Dr. Edmund Halley (for whom the famous comet is named) in 1690. Halley's diving bell was a large wooden cone. Lead weights on the cone were used to make it sink when placed in water. Divers could work underwater in the trapped air space. Diving bells were traditionally used for underwater observation or for recovering sunken wreckage. John Smeaton made the first modern diving bell, which was rectangular. Today's diving bells are much more complex than early designs. Today they are much like tiny submarines, with motors, supplemental air supplies, technology for removing carbon dioxide and replacing it with oxygen, and communication equipment.

Figure 3: A cutaway view of an early diving bell

REFERENCES

Chamber's Encyclopedia; Pergamon: Oxford, 1967; p 68.

Kaskel, A. *Principles of Science: Activity-Centered Program Teacher's Guide;*
    Charles E. Merrill: Columbus, OH; pp 29–30.

World Book Encyclopedia; World Book: Chicago, 1995; p 247.

CONTRIBUTORS

Lynda Dunlap, Southwestern Elementary School, Patriot, OH; Teaching Science
    with TOYS, 1991.

Susy Hasecoster, Liberty Elementary School, Liberty, IN; Teaching Science with
    TOYS, 1991.

Diana James, Farmers Elementary School, Farmers, KY; Teaching Science with
    TOYS, 1991.

Jerry Oberhaus, Liberty Center Elementary School, Liberty Center, OH; Teaching
    Science with TOYS, 1995.

Carole Pope, Southwestern Elementary School, Patriot, OH; Teaching Science with
    TOYS, 1991.

Jackie Toombs, Norwood Middle School, Norwood, OH; Teaching Science with
    TOYS, 1992–93.

Ann Veith, Rosedale Elementary School, Middletown, OH; Teaching Science with
    TOYS, 1991–92.

HANDOUT MASTER

A master for the following handout is provided:
• Instruction and Observation Sheet
Copy as needed for classroom use.

Tissue in a Cup
Instruction and Observation Sheet

1. Crush tissues and fit them tightly into the bottom of the plastic cup so that they won't fall out when the cup is turned upside down. You may need to use tape to hold the tissue in place, but be sure the tape does not cover the hole in the bottom of the cup.

2. Predict what will happen when you submerge the inverted cup containing the tissue in water, keeping your finger over the hole.

   _____

   _____

   _____

3. Put your finger firmly over the hole in the bottom of the cup. With the cup upside down, push the cup down into a container of colored water until it is completely submerged. (See figure.) You must keep the cup very straight as you push.

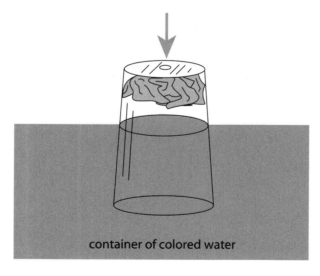

container of colored water

Keep the cup straight as you push it under water.

4. Keeping your finger on the hole, withdraw the plastic cup from the water. Remember to keep the cup straight as you lift it.

5. Dry the lip and outside of the cup with a paper towel and remove the tissue from the cup.

6. Feel the tissue and record your observations.

_____

_____

_____

7. Leaving your finger off the hole on the bottom of the cup, repeat Steps 1–6. Predict what will happen and record your observations. If the hole in the cup is small, you may need to submerge it a while before something happens.

Predictions: _____

_____

_____

Observations: _____

_____

_____

Showing That Air Has Mass

...Students use a soccer ball to discover that air meets the criteria for being matter: it has mass and takes up space.

✔ Time Required

Setup	none
Performance	10 minutes
Cleanup	none

Air trapped inside a soccer ball

✔ Key Science Topics

- criteria for matter
- gases
- mass

✔ Student Background

Students should understand that the two criteria for matter are having mass and occupying volume. Do at least one of the activities showing that air takes up space ("Balloon in a Bottle," "Burping Bottle," and "Tissue in a Cup") before doing this activity.

✔ National Science Education Standards

Science as Inquiry Standards:

- Abilities Necessary to Do Scientific Inquiry

  Students think critically and logically about their observations of the soccer ball and form a logical argument about the cause-and-effect relationship in the investigation.

Physical Science Standards:

- Properties and Changes of Properties in Matter

  The characteristic properties of all matter are having mass and taking up space.

✔ Additional Process Skills

• observing	Students see how pumping air into the ball affects its mass.
• measuring	Students determine the mass of the ball with various amounts of air in it.
• inferring	Students use the observed masses to conclude that air satisfies one of the criteria of matter.

MATERIALS

For the Procedure

Per class

- soccer ball or other semi-rigid, constant volume, yet inflatable ball
- air pump
- pump needle
- base from a plastic 2-L soft-drink bottle or other object to hold the ball in place on the balance pan
- balance (with capacity to measure the mass of the soccer ball, about 320 g, to at least the nearest 0.1 g)

 A double-pan balance can be used to show changes in mass without taking quantitative measurements.

SAFETY AND DISPOSAL

Do not over-inflate (and hence rupture) the ball. No special disposal procedures are required.

GETTING READY

If the ball is inflated to its normal playing pressure, insert the needle to equalize the inside and outside pressures, and then remove the needle. So that the ball will occupy its normal volume, do not squeeze the ball when removing the needle.

INTRODUCING THE ACTIVITY

Discuss and compare some of the properties characteristic of materials in the solid, liquid, and gaseous states. Ask if these materials are all forms of matter. Ask, "What is matter?" *Anything that has mass and occupies space is matter.*

Ask the students if they think that air is matter. What must they show in order to prove that air is matter? *That air has mass and occupies space.* Ask students how they would demonstrate that air takes up space. (The "Balloon in a Bottle," "Burping Bottle," and "Tissue in a Cup" activities in this book demonstrate that air occupies space. If students have done one or more of these activities, they should have plenty of suggestions.) Ask students how they would demonstrate that air has mass. Give students a chance to suggest ideas. Discuss the difficulty of measuring the mass of air. Then tell students that you have a way to demonstrate that air has mass.

PROCEDURE

1. Show students the soccer ball. Ask them what is inside the ball. *Air.* Place the soccer ball on the soft-drink bottle base in order to keep it from rolling off the balance. Ask a student volunteer to determine and record the combined mass of the ball and base. Ask, "What items does this mass measurement include?" *The ball, the base, and the air inside the ball.*

2. Remove the ball from the balance and bottle base and inflate the ball to normal playing pressure. Be careful not to overinflate (and hence rupture) the ball. Ask, "What have I just put inside the ball?" *More air.*

3. Ask students to predict whether the mass measurement of the ball will be different. Have them explain their predictions. Again, have a student volunteer measure the combined mass of the ball and the soft-drink bottle base.

4. As a class, compare the masses from Steps 1 and 3.

5. Ask students what conclusions they can make based on the data. If necessary, lead students to infer that air has mass, a necessary criterion for matter.

6. Remind students that they learned that air occupies volume in the activities "Balloon in a Bottle," "Burping Bottle," and "Tissue in a Cup." Ask, "What can you now conclude about air based on the results of the two activities?" *Since air occupies volume and has mass, air is matter.*

VARIATION

Set up the activity in a learning center and have students conduct the investigation using the Instruction and Observation Sheet (provided).

EXPLANATION

The following explanation is intended for the teacher's information. Modify the explanation for students as required.

Matter is anything that has mass and occupies volume. Both of these criteria are relatively easy to observe for liquid and solid samples but much more difficult for gaseous samples. This activity demonstrates that air (in the gas state) has mass, one of the two criteria of matter. (The "Balloon in a Bottle," "Burping Bottle," and "Tissue in a Cup" activities in this book demonstrate that air occupies space.)

This activity uses a ball with a semi-rigid case of leather or vinyl as a container for air. If the walls are not pushed in prior to adding extra air, such a container maintains a constant volume as more air is added. (Placing a sample of gas in a container that gets bigger when the gas is added, such as a balloon, does demonstrate that air occupies volume, but measuring the mass cannot be done directly in such an activity because of buoyancy effects. Buoyancy effects influence mass measurement because larger objects displace more air than smaller ones and so appear to have a smaller-than-actual mass.) The semi-rigid ball used in this activity does not significantly change volume when air is added; therefore, there is no visual evidence of additional air occupying additional volume, but accurate mass measurements clearly show an increase in mass.

ASSESSMENT

Inflate a soccer ball to its normal playing pressure. Ask the students to predict what will happen to the mass if the needle is inserted into the ball, creating a path for air movement in both directions. Do the experiment, finding the mass before and after inserting the needle. Ask students to explain their observations with pictures and words.

REFERENCE

Liem, T. "The Ball that Gains Weight;" *Invitations to Science Inquiry,* 2nd ed.; Ginn: Lexington, MS, 1987; p 42.

CONTRIBUTOR

John Williams, Associate Professor of Chemistry, Miami University Hamilton, Hamilton, OH.

HANDOUT MASTER

A master for the following handout is provided:
•   Instruction and Observation Sheet
Copy as needed for classroom use.

Showing That Air Has Mass
Instruction and Observation Sheet

1. Observe the soccer ball. What is inside it? _____

    Place the soccer ball on the soft-drink bottle base in order to keep it from rolling off the balance. Determine and record the combined mass of the ball and base. _____

    What items does this mass measurement include? _____

2. Remove the ball from the balance and bottle base and inflate the ball to normal playing pressure. Be careful not to overinflate (and hence rupture) the ball. What did you just put inside the ball?

    _____

3. Predict whether the mass of the ball will be different. Explain your prediction.

    _____

    _____

    Again, measure the combined mass of the ball and the soft-drink bottle base and record the mass.

    _____

4. Compare the masses from Steps 1 and 3. Was your prediction right? What conclusions can you make based on your data?

    _____

    _____

5. What are the two criteria for something to be classified as matter?

    _____

    _____

6. Remember that you have already learned that air takes up space in another activity. What can you now conclude about air based on the results of this activity and the fact that air takes up space?

    _____

    _____

    _____

Marshmallow in a Syringe

...Students use marshmallows to study the effect that changing pressure has on the volume of a gas.

✔ Time Required

Setup	5 minutes
Performance	15 minutes
Cleanup	5 minutes

Marshmallows in syringes

✔ Key Science Topics

- colloids
- gases
- compressibility of gases
- pressure-volume relationship of a gas (Boyle's law)

✔ Student Background

This activity may be used either to introduce and stimulate discussion of the pressure-volume relationship of gases or as an opportunity for students to apply their knowledge to explain the phenomenon of the expanding marshmallow. The activity may also be used to illustrate a common colloid.

✔ National Science Education Standards

Science as Inquiry Standards:

- Abilities Necessary to Do Scientific Inquiry

  Students think critically about the behavior of the marshmallow in the syringe and form a logical argument about the cause-and-effect relationship between gas pressure and volume.

Physical Science Standards:

- Properties and Changes of Properties in Matter

  A characteristic property of gases is that (for a given amount at constant temperature) their volumes vary inversely with pressure.

✔ Additional Process Skills

- inferring — Students develop ideas concerning the relationship between gas volume and pressure.

- observing — Students observe the volume of marshmallows at different pressures.

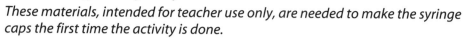

MATERIALS

For Getting Ready only

These materials, intended for teacher use only, are needed to make the syringe caps the first time the activity is done.

Per class
- Bunsen burner or candle
- 2 pairs of forceps, tweezers, or pliers
- small container of water
- disposable syringe needles (1 for each group)

Syringe needles are available from Fisher Scientific (#14-826-5B), 1600 W. Glenlake Ave., Itasca, IL 60143, 800/766-7000. Students will NOT be using the needle in this activity; they will use only the plastic connector.

For the Procedure
Per group
- several fresh miniature marshmallows of the same size
- syringe cap (made from the plastic connector of a disposable syringe needle; see Getting Ready)
- disposable plastic syringe with Luer-Lok tip

Syringes with capacities between 10 and 60 mL can be used. Several sizes of syringes are available from Fisher Scientific (60 mL #14-823-2D); syringes might also be available from a veterinarian or a farm supply store.

SAFETY AND DISPOSAL

Prepare the syringe cap (see Getting Ready) outside of class. Take care not to get melted plastic on your skin. Proper fire safety should be exercised, such as working on a fire-resistant surface and removing unnecessary flammable materials from the area. Long-haired people should tie hair back when working near flame. Dispose of the needle by placing it in the plastic needle cover and wrapping tape over the open end. Place the wrapped needle cover in the trash.

GETTING READY

Take care not to get melted plastic on your skin.

1. Prepare an inexpensive cap for each syringe as follows:

   a. Hold a disposable syringe needle with a pair of pliers, forceps, or tweezers, and use a candle or Bunsen burner to heat the needle close to where it enters the plastic connector.

   b. As the plastic begins to melt, pull the needle out with the second pair of pliers. (See Figure 1.) Drop the hot needle into the small container of water to cool.

c. After the plastic has cooled, check the cap for leaks as follows: Place the cap on the syringe with the plunger completely in the "in" position. Pull the barrel until it is about halfway out and then release the barrel. If it returns to its original position, the syringe cap is properly sealed. If the barrel does not return to its original position, some air is in the syringe, and the cap leaks. Try heating the plastic tip again to get it to seal or discard the cap and try again.

If the plastic catches fire, extinguish the fire by dipping the cap in water. Once prepared, the cap may be saved and used again. Dispose of the needle as specified in Safety and Disposal.

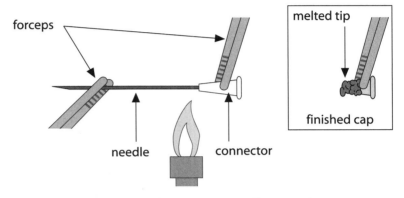

Figure 1: Remove the syringe needle to make a cap.

INTRODUCING THE ACTIVITY

Options:

- Do the "Properties of Matter" activity in this book, which demonstrates the compressibility of solids, liquids, and gases.

- Demonstrate (or have a student volunteer demonstrate) the behavior of gas in a syringe. First, with the plunger pushed in completely, try pulling out and pushing in the plunger, both with the cap on and with the cap off. (See Figure 2a.) Second, with some trapped air in the capped syringe, try pulling out the plunger. Repeat with the cap removed. (See Figure 2b.) Third, with some trapped air in the capped syringe, try pushing in the plunger. Repeat with the cap removed. (See Figure 2c.) Compare the results.

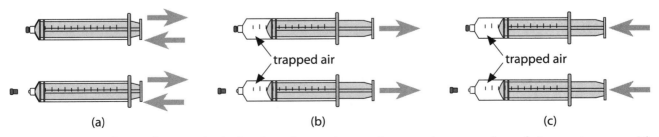

Figure 2: Have students observe the behavior of a gas in a syringe under a number of circumstances with the cap on (top) and off (bottom): (a) with the plunger completely in, pull the plunger out and push it back in; (b) with some air in the syringe, pull the plunger out; and (c) with some air in the syringe, push the plunger in.

PROCEDURE

Have student groups do the following:

If the syringe was filled with water in Introducing the Activity, it must be dried before putting the marshmallow in it to prevent the marshmallow from sticking to the sides of the syringe.

1. With the cap off, remove the plunger from the syringe barrel. Place one miniature marshmallow inside the syringe, leaving the cap off. Choose a second marshmallow of the same size to serve as the control.

2. Place the plunger in the syringe barrel and push it in to force out as much air as possible without squeezing the marshmallow.

3. Place the cap on the tip of the syringe.

4. Pull on the plunger, hold it in the "out" position and observe. (The marshmallow expands.)

5. Compare the size of this expanded marshmallow with the size of the control marshmallow.

6. Predict what will happen if the plunger is released. Release the plunger and observe.

7. Repeat Steps 4–6 several times, allowing each student within the group an opportunity to manipulate the syringe. Lead a class discussion of why the marshmallow behaves as it does, bringing out the relationship between gas volume and pressure. Encourage students to use their observations to support their ideas.

8. Remove the marshmallow from the syringe and compare it to the control marshmallow. (The marshmallow from the syringe usually appears shrunken and shriveled when compared to the control marshmallow.) Discuss why the two marshmallows are different. (See Explanation.)

VARIATIONS

1. Set up the activity in a learning center and have students conduct the investigation using the Instruction and Observation Sheet (provided).

2. For younger grades, have the students draw faces on the marshmallows before doing the Procedure. These faces make the growth and wrinkling more observable and fun.

EXPLANATION

The following explanation is intended for the teacher's information. Modify the explanation for students as required.

In this activity, the compressibility of a gas is illustrated. As the students observe when manipulating the syringe in Introducing the Activity, the air in the syringe is compressed when the plunger is pushed, and it expands when the plunger is pulled out. The behavior of a gas in a sealed syringe is in contrast to the behavior of water in the sealed syringe when the plunger is pushed and pulled (because liquids are essentially incompressible).

This activity also illustrates the pressure-volume relationship of a gas, known as Boyle's law. Boyle's law states that volume varies inversely with pressure at constant temperature and constant amount of gas; therefore, as volume increases, pressure decreases. Likewise, if volume decreases, pressure increases. In the closed system of the capped syringe, when the plunger is pulled out, volume increases and the pressure inside the syringe decreases.

The marshmallow is a convenient item to illustrate Boyle's law because it is a gas in a solid, one type of colloid. (Similar colloids are whipped cream and shaving lather.) When the pressure inside the syringe is reduced by pulling out the plunger, the volume of trapped air in the marshmallow expands, and the volume of the marshmallow increases. As the plunger is released, volume decreases, pressure increases, and the process is reversed. The marshmallows often appear shrunken afterwards because some of the air initially trapped inside them escapes when the pressure is decreased.

ASSESSMENT

Have students prepare data sheets similar to Table 1. Tell students that their data sheet responses should be descriptors such as "same," "less," and "more."

Ask students to imagine a marshmallow sealed on Earth in a steel container at 1 atmosphere pressure. Ask students to predict what change (if any) would occur in the volume of a marshmallow if the container were opened in each of the following situations (assuming no temperature change):

1. in the space shuttle ("weightless" but pressurized at 1 atmosphere),
2. on a space walk ("weightless" with zero atmospheric pressure), and
3. on a deep-sea dive at a depth of 100 feet. (The pressure is about 4 atmospheres.)

Table 1: Sample Student Data Sheet		
	Volume Before Opening	Volume After Opening
Inside a Space Shuttle		
During a Space Walk		
During a Deep-Sea Dive		

Answers:
1. The volume of the marshmallow inside the space shuttle will be the same as on Earth.
2. Before opening the container on a space walk, the marshmallow's volume will be the same as on Earth; when the container is opened on the space walk, the marshmallow will expand, and some trapped gas will escape.
3. Before opening, the volume of the marshmallow will be the same as at sea level; when the container is opened, the marshmallow will decrease in volume (due to higher pressure).

CROSS-CURRICULAR INTEGRATION

Earth science:
- Mark a map of the U.S. or the world with one color for high-altitude areas which would have lower atmospheric pressure. (You decide the cutoff altitude.) Discuss the effects of this lower pressure, such as increased baking time and rapid fatigue during exercise in people who are not accustomed to high altitudes.

Language arts:
- Have students read one or more short stories about space travel by Isaac Asimov. (*I Robot* is a good collection.) Ask them to look for any references to pressure changes (such as use of air locks and space suits, illnesses due to space travel).

Life science:
- Have students investigate the physiological adaptations of Korean pearl divers and of deep-sea organisms that live under great pressure.
- Have students investigate the physiological adaptations of people who live at very high altitudes or athletes who train at high elevations.

Math:
- Have students mathematically determine the volume of the miniature marshmallow 1) before it is expanded and 2) after it has been expanded and removed from the syringe. (The volume of a cylinder is $\pi r^2 h$ where r is the radius and h is the height of the cylinder.) Discuss the difference in volume and the reason for the difference.

CONTRIBUTORS

Alison Dowd, Talawanda Middle School, Oxford, OH; Teaching Science with TOYS peer writer.

Gary Duncan, Harvard Elementary School, Toledo, OH; Teaching Science with TOYS, 1992.

Steve Peterson, Burroughs Elementary School, Toledo, OH; Teaching Science with TOYS, 1992.

HANDOUT MASTER

A master for the following handout is provided:

*   Instruction and Observation Sheet

Copy as needed for classroom use.

Marshmallow in a Syringe
Instruction and Observation Sheet

1. With the cap off, remove the plunger from the syringe barrel. Place one miniature marshmallow inside the syringe, leaving the cap off. Choose a second marshmallow of the same size to serve as the control.

2. Place the plunger in the syringe barrel and push it in to force out as much air as possible without squeezing the marshmallow.

3. Place the cap on the tip of the syringe.

4. Pull on the plunger, hold it in the "out" position and observe. Record your observations.

   _____

   _____

5. Compare the size of this expanded marshmallow with the size of the control marshmallow.

   _____

   _____

6. Predict what will happen if the plunger is released.

   _____

   _____

7. Release the plunger. Record your observations.

   _____

   _____

8. Remove the marshmallow from the syringe and compare it to the control marshmallow. Why do the two marshmallows look different?

   _____

   _____

Moving Molecules

...Students conclude that temperature affects the rate of movement of molecules in liquids and gases.

✔ Time Required

Setup   5 minutes
Performance   10–15 minutes for Part A
Cleanup   5 minutes

Food color in water

✔ Key Science Topics

- diffusion
- energy
- matter
- molecules

✔ National Science Education Standards

Science as Inquiry Standards:

- Abilities Necessary to Do Scientific Inquiry

  Students develop descriptions, explanations, predictions, and models using evidence from the investigation.

Physical Science Standards:

- Properties and Changes of Properties in Matter

  A characteristic property of liquids and gases is that the particles do not have a fixed arrangement.

✔ Additional Process Skills

• observing	Students observe the swirling action of the food color in water and the diffusion of scent through a room.
• communicating	Students describe their observations.

MATERIALS

For Getting Ready
Per class or group
- 2 clear plastic cups
- felt-tipped pen
- masking tape

For Introducing the Activity
Per class
- clear plastic cup filled with room-temperature water
- food color

For the Procedure
Part A, per class or group
- labeled cup filled with hot water (See Getting Ready.)
- labeled cup filled with cold water (See Getting Ready.)
- 2 clear plastic cups filled with same-temperature water
- food color
- spoon

Part B, per class
- (optional) Fisher-Price™ Magic MiniVac or "popcorn popper" push toy

For Variations and Extensions
❶ All materials listed for the Procedure except
Per class or group
- substitute tea bags for the food color

❷ All materials listed for the Procedure plus the following:
Per class
- ice
- heat-safe container for heating water on a hot plate
- hot plate
- hot-beverage cup
- alcohol thermometer
- stopwatch

SAFETY AND DISPOSAL

No special safety or disposal procedures are required.

GETTING READY

Use masking tape and a felt-tipped pen to label the cups containing the hot and cold water.

Investigating Solids, Liquids, and Gases with **TOYS**

INTRODUCING THE ACTIVITY

Show the students the bottle of food color and a cup of water. Ask them to predict what would happen if you added a drop of food color to the water. Have the students observe as you add a drop of food color to the water. Now ask them to give you ideas on what could be done to speed up the mixing of the food color and the water. Use the students' ideas to lead into the activity.

PROCEDURE

Part A: Diffusion in Liquids

1. Add a drop of food color to each of two cups of same-temperature water. Use a spoon to stir the contents of one of the cups but not the other. Discuss the results and have the students compare these to their predictions from Introducing the Activity.

2. Show the students the labeled cups of hot and cold water and ask them to predict how the food color will behave when added to these cups.

3. Add one drop of food color to each labeled cup of water. Do not move the cups or stir the water.

4. Have the students watch the areas of color in each cup and describe what is happening. Let them offer possible explanations for their observations. Test the explanations when possible. For example, if you used different food colors and students suggest that different colors disperse at different rates, try the investigation again using the same color in both cups.

Part B: A Model

1. Have students develop a general statement regarding rate of movement versus temperature.

2. (optional) Show the students the MiniVac or "popcorn popper" push toy. (See Figure 1.) The toy can be used to simulate the rate of motion of molecules in liquids or gases. The motion of the balls when the toy is pushed slowly is analogous to the movement of molecules in a colder system. When the toy is pushed more rapidly, the motion of the balls is analogous to the movement of molecules in a warmer system.

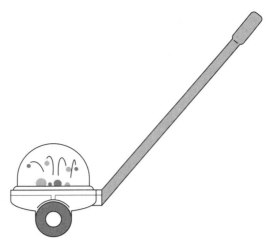

Figure 1: A Popcorn Popper Toy

VARIATIONS AND EXTENSIONS

1. Use tea bags instead of food color.

2. For older students, more varied temperatures can be used in Part A. For example, you could have students try ice water, room-temperature water, hot tap water, and water heated to nearly boiling on a hot plate. Students can measure the temperature of each type of water with a thermometer and use a stopwatch to measure the time required for complete mixing. Final results can be graphed.

 Students will have to do some preliminary trials so they can decide how to determine when mixing is complete.

EXPLANATION

The following explanation is intended for the teacher's information. Modify the explanation for students as required.

The particles of both liquids and gases are in constant motion and do not have a fixed arrangement, so the particles can change neighbors. The molecules of water in the hot water have greater energy than those in cold water. This greater energy causes the molecules in the hot water to move at a faster rate than those in the colder environment.

While the movement of molecules is not visible, we can observe the effect of this movement. In the activity, the food color is added to both hot and cold water. Although it is not stirred, the food color spreads out through the water. This type of movement is called diffusion. The colored molecules diffuse more rapidly in the hot water than in the cold water because of the faster rate of movement of the particles in the hotter water.

REFERENCES

Mebane, R.; Rybolt T. *Adventures with Atoms and Molecules;* Enslow: Hillside, NJ, 1985; pp 10–11.

The Raintree Illustrated Science Encyclopedia; Raintree: Milwaukee, WI, 1984; Vol. 12, p 1093.

The World Book Encyclopedia; World Book: Chicago, 1988; Vol. 13, p 697.

CONTRIBUTORS

Joyce Cook, Northwestern Elementary School, Kokomo, IN; Teaching Science with TOYS, 1992–93.

Susan Higgins, Union Elementary School, Biggsville, IL; Teaching Science with TOYS, 1995.

Marcia Jones, Springfield City Schools, Springfield, OH; Teaching Science with TOYS, 1991–92.

Barbara Yochum, Concord Elementary, Hillsboro, OH; Teaching Science with TOYS, 1989–90.

Non-Newtonian Fluids—Liquids or Solids?

...Not all matter is easily classified as a solid, liquid, or gas.

✔ Time Required

Setup	60–75	minutes
Performance	30–45	minutes
Cleanup	10–15	minutes

✔ Key Science Topics

- liquids
- non-Newtonian fluids
- solids

✔ National Science Education Standards

Science as Inquiry Standards:

- Abilities Necessary to Do Scientific Inquiry

   Students make systematic observations of four non-Newtonian fluids.

   Students gather and organize data.

   Students communicate observations.

Physical Science Standards:

- Properties and Changes of Properties in Matter

   The characteristics of non-Newtonian fluids make them difficult to classify as liquids or solids.

✔ Additional Process Skill

- predicting   Students predict the outcome of a "race" between the various non-Newtonian fluids.

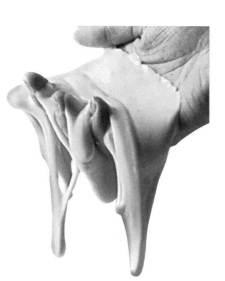

Gluep is a non-Newtonian fluid.

MATERIALS

For Introducing the Activity
- Slime
- Ooze
- Silly Putty®

For Getting Ready
For Gluep, per class
- 30 mL (2 tablespoons) laundry borax (sodium tetraborate decahydrate, $Na_2B_4O_7 \cdot 10H_2O$)
- (optional) Lysol® Deodorizing Cleaner
- 500 mL (about 2 cups) white glue (such as Elmer's Glue®)

➤ *While not common, a few types of glue produce more watery Gluep than desired. Test your glue by carrying out a single preparation in advance. If the glue-water mixture does not clump up or adhere to the stick when stirred, you should add a bit more glue in the glue-water mixture and repeat the test.*

For Slime, per class
- 15 mL (1 tablespoon) laundry borax
- (optional) Lysol® Deodorizing Cleaner
- polyvinyl alcohol solution (enough for 30 mL per student), either purchased (see note below) or made using the following:
  - 40 g (just over ⅓ cup) polyvinyl alcohol powder
  - water
  - 2-L beaker or 2-quart heatable container
  - hot plate/magnetic stirrer combination OR hot plate and stirring rod
  - volume measuring devices such as 10- and 100-mL graduated cylinders or teaspoon, tablespoon, and cup measures
  - (optional) about 5 mL (1 teaspoon) solid sodium hydroxide (NaOH)

➤ *The polyvinyl alcohol must be 99% or 100% hydrolyzed with a molecular weight of at least 100,000. It is available in powder form (#P0153 for 100 g) or as a 4% aqueous solution (#P0209 for 500 mL from Flinn Scientific, P.O. Box 219, Batavia, IL 60510-0219; 800/452-1261).*

For the Procedure
Per class or group
- ring stand with ring clamp
- water-soluble marker
- stopwatch
- bowl
- wide-stemmed funnel

➤ *The cut-off top of a 1- or 2-liter (L) plastic bottle works well. (See Figure 1.)*

For all parts, per student
- goggles

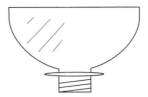

*Figure 1: Make a wide-stemmed funnel
from the cut-off top of a 1- or 2-L plastic bottle.*

Corn Starch Putty, per student
- aluminum pie pan or other shallow container
- ¼ cup corn starch
- 1–2 tablespoons water
- (optional) food color

Gluep, per class
- 1 of the following sets of measures:
  - tablespoon, teaspoon, and cup measures
  - 10-, 50-, and 100- to 500-mL graduated cylinders
- water
- (optional) food color

Gluep, per student
- 10 mL (2 teaspoons) borax solution prepared in Getting Ready
- 30 mL (2 tablespoons) glue-water solution prepared in Getting Ready
- craft stick, ice cream stick, or tongue depressor
- 2 disposable 3- to 6-ounce paper or plastic cups
- zipper-type sandwich bag

Slime, per student
- 20 mL 4% aqueous solution of polyvinyl alcohol (purchased or made in Getting Ready)
- 2–3 mL (½ teaspoon) borax solution prepared in Getting Ready
- 5-ounce plastic or paper cup
- craft stick, ice cream stick, or tongue depressor
- (optional) 1–3 drops food color
- (optional) zipper-type plastic bag

Laundry Starch/Glue Putty, per student
- 30 mL (2 tablespoons) white glue
- 15 mL (1 tablespoon) Purex® Sta-flo liquid laundry starch
- 5-ounce paper cup
- plastic spoons or stirring stick
- measuring spoons or graduated cylinder
- zipper-type plastic bag
- index card and water-soluble marker
- (optional) food color

SAFETY AND DISPOSAL

Some people have an allergic reaction to dry, powdered sodium tetraborate (borax). Use caution when handling it. Avoid inhalation and ingestion. Use adequate ventilation when preparing the borax solution, and wash your hands after contact with the solid.

Typically no danger is involved in handling Gluep or Slime, but have students wash their hands after use. Persons with especially sensitive skin or persons who know they are allergic to borax or detergent products should determine their sensitivity to Gluep or Slime by touching a small amount. Should itching or redness occur, wash the area with a mild soap and avoid further contact with the Gluep or Slime.

Gluep does not readily stick to clothes, walls, desks, or carpet. Caution students not to place Gluep on wood furniture, since it will leave a water mark. If Slime spills on the carpet, apply vinegar to the spot and follow with a soap-and-water rinse. Do not let the Slime harden on carpeting. Do not put Slime on wood furniture; it will leave a water mark.

Gluep and Slime will keep a long time if stored in a plastic bag. To retard the growth of mold in Gluep and Slime, add a few drops of Lysol Deodorizing Cleaner to the borax solution used to make them. If you use Lysol, follow all precautionary statements on that product's label.

If sodium hydroxide is used to remove the polyvinyl alcohol residue from the container in which you prepared the 4% solution, exercise extreme caution and wear eye protection; sodium hydroxide is very caustic. If contact with skin or eyes occurs, rinse with water for 15 minutes; if contact is with the eyes, seek medical attention while rinsing is occurring.

If you allow students to take any of the non-Newtonian fluids home, send a warning not to give them to small children. Also, send the precautions and cleanup instructions from the preceding paragraphs.

Solutions prepared in Getting Ready can be stored for several months. Shake well before use if stored for long periods of time. Discard Gluep in a waste can. Discard Slime in a waste can or flush it down the drain with lots of water. Discard Corn Starch Putty in a waste can.

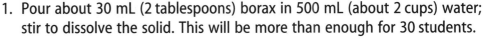

GETTING READY

For Making Gluep

1. Pour about 30 mL (2 tablespoons) borax in 500 mL (about 2 cups) water; stir to dissolve the solid. This will be more than enough for 30 students.

➤ *Adding several drops of Lysol Deodorizing Cleaner to the borax solution can retard mold growth in the Gluep that can result from students handling the Gluep with dirty hands.*

2. Mix the white glue with an equal volume of water, allowing for about 2 tablespoons of the mixture per student. Example: 500 mL (about 2 cups) glue plus 500 mL (about 2 cups) water will give 1 L (about 4 cups) glue-water mixture, enough for 30 students. Stir or shake thoroughly until well mixed.

3. (optional) For younger students, you may wish to prepare one cup containing 30 mL (2 tablespoons) glue-water mixture and a second cup containing 10 mL (2 teaspoons) borax solution (see Step 1) for each student.

For Making Slime

1. Make a borax solution. Pour about 15 mL (1 tablespoon) borax in 250 mL (about 1 cup) water; stir to dissolve the solid.

➤ *Adding several drops of Lysol Deodorizing Cleaner to the borax solution can retard mold growth in the Slime that can result from students handling the Slime with dirty hands.*

2. If 4% polyvinyl alcohol solution is not purchased, make a polyvinyl alcohol solution according to one of the following procedures:

➤ *The polymer solution is clear and colorless. It has a syrupy texture and traps air bubbles when stirred. Freshly prepared solution is usually odorless, but as it ages, an odor can develop. A drop of food color can be added to the polymer solution to color the Slime, but the color may interfere with the process of lifting ink from paper.*

MICROWAVE METHOD

a. Dissolve 40 g (⅓ cup) polyvinyl alcohol powder in 1 L water in a microwave-safe container.

b. Stir the solution and place it in a full-size microwave. Heat the solution on high for 8 minutes, stirring every 1–2 minutes. Do not make more than 1 L at a time.

➤ *Heating time may vary depending on the power of your microwave oven.*

HOT PLATE METHOD

a. While stirring, sprinkle 40 g (⅓ cup) of polyvinyl alcohol into 1 L water.

b. Continually stir the mixture while heating it on a hot plate over a moderately high heat.

➤ *The solution will initially be quite milky in color, but it will clear when the polyvinyl alcohol is completely dissolved. The process may take up to 45 minutes. The mixture may be heated rapidly, but be careful not to scorch the syrupy liquid.*

c. Allow the solution to cool before using. If a slimy or gooey layer appears on the top upon cooling, skim it off and discard. The solution can be stored for several months in a sealed container.

➤ *If a solid residue remains in the glass container used for heating, it may be removed by peeling it off. If this is not possible, add 1 teaspoon sodium hydroxide (lye) and enough water to cover the residue, then allow it to sit until the solid residue is loosened. Do not add sodium hydroxide to a metal pot. Use extreme caution and eye protection when handling sodium hydroxide; it is very caustic. If contact with skin or eyes occurs, rinse with water for 15 minutes; if contact is with the eyes, seek medical attention while rinsing is occurring.*

INTRODUCING THE ACTIVITY

Show students several commercially available types of non-Newtonian fluids (such as Slime, Ooze, and Silly Putty). Allow students to examine them and attempt to classify them as a solid, liquid, or gas. Have them discuss the problems associated with such a classification. Have the students consider these problems as they prepare and observe the non-Newtonian fluids made in the activity.

PROCEDURE

➤ *You may wish to select only one of the following preparations or to assign a different preparation to each group and have the class compare properties of their products once completed.*

1. Have students make the non-Newtonian fluids as outlined on the Instruction Sheet (provided). Tell students to carefully observe each ingredient before and after mixing.

2. Discuss the importance of systematic observation in a scientific investigation. Ask students how they plan to record and organize their observations. Discuss different possibilities. Have students compare the properties of the fluids by squeezing each substance, forming it into a ball, and throwing it onto a tile or linoleum floor; by pulling it slowly and then quickly; and by pressing the fluid on top of a name written with a water-soluble marker. Separate the fluid by breaking it into pieces.

3. Have students share their observations of the fluids and consider the following questions: In what ways did they behave like ordinary fluids? In what ways were their behaviors unusual? Did all the fluids behave alike? If not, how were they different?

4. Explain to students that you will now hold a race between the fluids. Set up a ring stand, ring clamp, wide-mouthed funnel, and bowl as shown in Figure 2. Explain to students that you're going to place one of the non-Newtonian fluids in the funnel. Tell them that you'll measure and record the amount of time required for all of the fluid to run out of the funnel. Then you'll repeat the process with an equal volume of each of the other non-Newtonian fluids. Based on their observations from Steps 2–3, have students predict which non-Newtonian fluid will run out of the funnel the fastest, which is next fastest, and which is slowest and explain the reasons for their predictions. Place one of the non-Newtonian fluids in the funnel and begin timing. Stop timing when all of the substance is in the bowl. Repeat for the other non-Newtonian fluids.

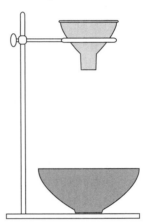

Figure 2: Set up a ring stand, ring clamp, wide-mouthed funnel, and bowl.

EXPLANATION

The following explanation is intended for the teacher's information. Modify the explanation for students as required.

Corn Starch Putty, Gluep, Slime, and Starch/Glue Putty belong to a class of materials that do not obey the usual laws of viscosity and are called non-Newtonian fluids. Viscosity, one of the physical properties of a liquid, describes resistance to flow. For example, water and alcohol are described as having low viscosity because they flow quickly; honey and syrup have high viscosities because they flow much more slowly. Most fluids are Newtonian; that is, their ability to be poured depends on temperature. In non-Newtonian fluids, variables besides temperature affect their ability to be poured.

A low stress, such as slow pulling, allows non-Newtonian fluids to flow and stretch and even form a thin film. A high stress, such as pulling sharply, will cause the material to break. Hitting a piece of Corn Starch Putty, Gluep, Slime, or Starch/Glue Putty with a light hammer will not cause splashing or splattering, and the material will bounce to a small extent. If pushed through a tube, the material will emerge with a swell (known as a die swell in the plastics extrusion trade).

Corn Starch Putty is a suspension of corn starch in water. It looks wet but becomes powdery when you rub it between your fingers. It withstands sudden shocks, but it doesn't support the weight of an object laid on its surface. This is a result of its dilatant nature; it tends to dilate (or expand) under stress. In other words, it tends to "push back" when struck with a firm blow. The viscosity of Corn Starch Putty increases when a stress such as rapid stirring is applied, but will pour freely when not under stress.

Gluep and Starch/Glue Putty contain polymers; polymers are molecules with long chains of repeating units—much like a chain of paper clips. The glue solution you use to make Gluep and Starch/Glue Putty contains millions of individual chains of a polymer called polyvinyl acetate, often used in latex paint. Before you add the borax, sodium tetraborate (or the laundry starch, which contains borax), these chains are able to slip and slide freely over one another like strands of freshly cooked spaghetti. When you add the borax (or laundry starch) solution to the glue, you cause the polymer chains to be linked together—just as the rungs link the two sides of a ladder. This process is called cross-linking. After the polymer becomes cross-linked, the individual chains can no longer slip and slide. In this process, the liquid glue mixture becomes the semisolid Gluep or Starch/Glue Putty.

The polyvinyl alcohol solution in the Slime recipe contains long polymer chains of polyvinyl alcohol dissolved in water. Because these chains are very long, they interfere with each other's movement, causing this solution to be rather viscous. The preparation of Slime is completed when the solution of cross-linker, borax, is added to the polyvinyl alcohol solution. The cross-linker bonds the different polyvinyl alcohol chains together. The resulting Slime is more viscous than the original polymer solution.

CROSS-CURRICULAR INTEGRATION
Math:
- Have students prepare a bar graph showing the results of the race between the fluids. How are the race times and the viscosities of the fluids related? Which fluid is most viscous? Least?

REFERENCES

Casassa, E.Z.; et al. "The Gelation of Polyvinyl Alcohol with Borax," *Journal of Chemical Education.* 1986, *63,* 57–59.

"Corn Starch Putty"; "Make-it-Yourself Slime"; *Fun with Chemistry: A Guidebook of K–12 Activities;* Sarquis, M., Sarquis, J., Eds.; Institute for Chemical Education: Madison, WI, 1991; Vol. 1, pp 39–42.

"Homemade 'Slimes'"; *Chain Gang—The Chemistry of Polymers;* Sarquis, M., Ed.; Terrific Science: Middletown, OH, 1995; pp 111–117.

Sarquis, A.M. "Dramatization of Polymeric Bonding Using Slime," *Journal of Chemical Education.* 1986, *63,* 60–61.

Stroebel, G.G.; et al. "Slime and Poly (Vinyl Alcohol) Fibers: An Improved Method," *Journal of Chemical Education.* 1993, *70*(40), 893.

HANDOUT MASTER

A master for the following handout is provided:

• Instruction Sheet

Copy as needed for classroom use.

Non-Newtonian Fluids—Liquids or Solids?

Instruction Sheet

Copy the following instructions and cut them apart at the dotted lines. Give copies to students to use for the hands-on activity.

Making the Corn Starch Putty

1. Place ¼ cup corn starch in the pie pan.

2. Using your fingers to mix, add water slowly until a gooey, fluid-like consistency is achieved. If desired, add food color to the water before mixing.

3. Observe the gooey fluid. The putty should pour or drip slowly from your hand, but when you strike it with a hard blow, it should not splatter. (See Figure.) If you added too much water, add additional corn starch to reach the desired consistency.

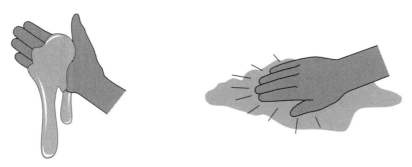

The Corn Starch Putty has the right consistency when it drips slowly from the hand but does not splatter when struck with a hard blow.

Making Gluep

1. Measure about 30 mL (2 tablespoons) glue-water mixture into a cup.

2. (optional) Add 1–2 drops food color to the glue-water mixture.

3. Stir the glue-water mixture with a stick and describe its properties.

4. Stir in about 10 mL (2 teaspoons) borax solution. Continue to stir until a mass of Gluep forms on the stir stick.

5. Store the Gluep in a zipper-type plastic bag. (Gluep will last a long time if not allowed to dry. It will eventually get moldy; discard it in a waste can.)

Reproduced from *Investigating Solids, Liquids, and Gases with* **TOYS**, published by Terrific Science Press.

Making Slime

1. Pour about 30 mL (2 tablespoons) of the 4% polyvinyl alcohol solution into a cup.

2. (optional) Add 1–2 drops food color to the polyvinyl alcohol solution.

3. Stir the solution with a stick.

4. While stirring, pour about 2–3 mL (½ teaspoon) borax solution into the cup of polyvinyl alcohol. Be sure to stir with the stick as the borax solution is added.

5. Once the gel has formed, remove the Slime from the cup and knead it in your hands. The gel will develop a consistency comparable to commercial Slime and other similar materials that are sold in toy stores.

6. Put the Slime in a zipper-type plastic bag for storage. (The Slime lasts 2 days to 2 weeks. It will eventually get moldy; discard it in a waste can.)

Making Laundry Starch/Glue Putty

1. Pour 30 mL (2 tablespoons) white glue into a paper cup.

2. (optional) Add 1–2 drops food color to the glue. Mix thoroughly.

3. Add 15 mL (1 tablespoon) of Sta-flo laundry starch while stirring. A thick mass will begin to form on the stirring stick. Continue to stir and work the glue throughout.

4. Reach in and remove the putty with your fingers. Form into a ball and rinse under running water.

5. Place the putty in a zipper-type plastic bag to prevent it from drying out.

☑ **Hands-On Activity**

☐ **Demonstration**

☐ **Learning Center**

Rock Candy Crystals

...Students observe crystals form as the solvent evaporates from a solution.

✔ Time Required

Setup   5   minutes
Performance 30–45 minutes (+ 1–2 weeks)
Cleanup   5–10 minutes

✔ Key Science Topics

- crystals, crystallization
- evaporation
- saturated solutions

✔ Student Background

Students should have some knowledge of evaporation and of the states of matter.

✔ National Science Education Standards

Science as Inquiry Standards:

- Abilities Necessary to Do Scientific Inquiry

  Students make systematic observations of sugar solutions as they watch rock candy form.

Physical Science Standards:

- Properties and Changes of Properties in Matter

  A characteristic property of most solids—including sugar—is that their solubility in water increases with heating. Cooling and evaporation result in the solute coming out of solution.

Rock candy

✔ Key Process Skills

- observing   Students observe the formation of solute crystals as the solvent evaporates.

- measuring   Students measure ingredients to make a saturated solution.

MATERIALS

For Introducing the Activity
Per class
- 1 of the following:
  - ○ jar of honey that has crystallized
  - ○ rock candy (available at candy stores)
- magnifying lens

For the Procedure
These materials must be food-safe (previously used only for food).

Per class
- mini muffin tins, enough to provide 1 muffin well for each student

Larger muffin tins or aluminum tart pans can be used, but these will require considerably more sugar solution.

- masking tape or labels to label toothpicks or strings
- measuring cups
- hot plate
- wooden spoon
- saucepan
- hot pads
- paper towels

Per group
- ¼ cup water
- ½ cup granulated sugar in a cup or other container

This makes enough solution to fill 5–6 mini muffin wells. To fill 4 regular-sized muffin wells, double the recipe.

- plastic cup (large enough to hold the sugar and the water)
- teaspoon measure
- spoon or craft stick
- toothpick or piece of string for each group member
- magnifying lens

For Variations and Extensions
The materials for Extensions 1 and 2 must be food-safe (previously used only for food).

❶ Per class
- 8 ounces pure maple syrup

Eight ounces of syrup will make about 20 1-inch candies.

- heavy saucepan
- hot plate
- candy thermometer

- wooden spoon
- plastic candy molds
- plate
- non-stick spray

❷ Per class
- 8 ounces pure maple syrup

Eight ounces of syrup will make about ¾ cup sugar.

- heavy saucepan
- hot plate
- candy thermometer
- non-stick spray
- cookie sheet
- metal spatula
- blender or food processor

❸ Per class
- broom handle or chair
- pieces of blue paper

SAFETY AND DISPOSAL

When the sugar-water solution is heated, it will become hot enough to cause burns if splattered or spilled on the skin. The teacher or another adult should heat and pour the solution and should take appropriate precautions.

Because the rock candy may be eaten, all the materials must be food-safe (meaning they must have been used previously only for food), and everyone involved should have clean hands. Cover the mini muffin tins with paper towels to keep foreign matter out but still allow evaporation to occur.

No special disposal procedures are required.

INTRODUCING THE ACTIVITY
Options:

- Have students examine a jar of honey that has crystallized. Lead a discussion by asking questions such as, "Has this ever happened to honey at your house?" "What do you see in the jar?" "What does it look like?" "Where did the crystals come from?" Give the students time to examine the sugar crystals in the honey with a magnifying lens. Explain that crystals are solids that have regular geometric shapes.

- Give each group or each student a piece of rock candy. Have the students examine the candy with a magnifying lens and describe what they observe. Explain that crystals are solids that have regular geometric shapes.

PROCEDURE

1. Have the students wash their hands.

2. Have each student write his or her name on a small piece of masking tape or a small label. Provide each student with a toothpick or piece of string and have each student stick the label on one end as shown in Figure 1.

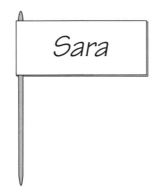

Figure 1: Fold a piece of masking tape over a toothpick or string to make a label.

3. Have each group of students gather ¼ cup water in a plastic cup, ½ cup granulated sugar in a cup, a teaspoon measure, and a spoon or craft stick.

4. Have the students guess how many teaspoons of sugar they will be able to dissolve in their ¼ cup water. The students should then add 1 teaspoon sugar to the water. Have them use the second spoon or craft stick to stir the solution until all of the solid dissolves. They should continue adding sugar 1 teaspoon at a time, keeping track of the number of teaspoons of sugar added, until the sugar stops dissolving. When the sugar solution is saturated, undissolved sugar remains, even after 2–3 minutes of stirring. Collect the sugar that each group did not add to the solution—you'll use it in Step 6.

5. Explain to the students that they have made a saturated solution. Have each group share with the class how many teaspoons of sugar they used and suggest reasons why they may not have all used the same amount of sugar. *Suggestions might include that some groups filled their spoons more full, some groups did not stir long enough after adding sugar to see that it would still dissolve, etc.* Also discuss how their guesses compared with the actual number of teaspoons of sugar.

6. Tell the students that you would like to dissolve more sugar in the same amount of water and ask for suggestions as to how you might do this. *Heating the solution will allow more sugar to dissolve.* Combine the solutions from each group into one saucepan. Add the leftover sugar from each group. Heat while stirring until all of the sugar has dissolved. Bring to a boil and turn down the heat. Boil while stirring for about 1 minute. The solution should be thick and clear and contain no sugar crystals.

The sugar solution is very hot, so use appropriate caution.

7. Immediately pour the solution into the mini muffin cups.

Caution the students not to touch the hot muffin tins.

8. Have each student insert a labelled toothpick or piece of string into a separate mini muffin well.

9. After the solution cools, place the muffin tins where they will not be disturbed and cover them with paper towels to keep foreign matter out but still allow evaporation to occur.

10. It will take a week or two for the water to evaporate. Have students record daily observations on the Data Sheet (provided). If a crust forms on the surface, break it with a clean spoon or fork and carefully remove it so the water can continue to evaporate.

11. After the crystals have formed on the toothpicks, have the students remove the toothpicks with attached crystals from the mini muffin cups. Have students examine the crystals using a magnifying lens and record their observations. Allow the students to eat their crystals.

VARIATIONS AND EXTENSIONS

The maple syrup solution is very hot, so use appropriate caution. Use only food-safe containers for Extensions 1 and 2.

1. Make maple sugar candy from maple syrup as follows: Lightly coat plastic candy molds with non-stick spray. Heat 8 ounces maple syrup over high heat until boiling. Turn heat down to medium-low—just high enough to keep the syrup boiling. Do not stir. Boil until the syrup reaches 244°F (118°C). Take the syrup off the heat and carefully run warm water along the side of pot to cool the syrup. (Take care not to run water *into* the pot as you do this.) Stir the syrup as it cools, watching for the syrup to lighten in color and begin to thicken. Once this happens, work quickly to pour syrup into molds because the syrup will very quickly become too thick to pour. (If the syrup does thicken too much, scrape it into a microwave-safe container, microwave for 20–30 seconds on full power, and then pour it into the molds.) When the candy is firm, unmold it onto a plate. Have students examine the candy carefully for crystals. Then eat!

2. Make maple sugar from maple syrup as follows: Lightly coat a cookie sheet with non-stick spray. Heat 8 ounces maple syrup to boiling on high heat. Reduce heat to medium-low and continue to boil until temperature reaches 250°F (121°C). Do not stir. At this point, the syrup should appear thicker, and bubbles should rise and pop a bit more slowly. Pour the hot syrup onto the cookie sheet and begin scooping and turning the syrup with a spatula. As the syrup cools it will lighten, thicken, and eventually harden. Keep breaking up the hardening syrup until the largest pieces are no more than ¼ inch in diameter. When cool, blend the sugar in a blender or food processor to a fairly uniform consistency. Allow students to examine the maple sugar for crystals and to have a taste.

3. As a kinesthetic demonstration, give each student a piece of blue paper to represent water and tell students that they represent the sugar dissolved in water. Use a broom handle or similar object or a chair to represent the toothpick. Have the students move around the room holding their "water." When a student comes close to the broom, place his or her "water" on the floor and have the student remain still, holding the broom. When a second student passes close to the broom, repeat the removal of the water and have the second student hold hands with the first student. Continue the process, placing the students in a regular, orderly arrangement until most are part of the crystal growing off the broom. Discuss the crystallization process based on the model.

4. Have students do the Take-Home Activity (handout provided) outside of school with an adult partner.

EXPLANATION

The following explanation is intended for the teacher's information. Modify the explanation for students as required.

Crystals are solids that have regular geometric shapes. This shape is the result of a regular arrangement of particles that make up the crystals. A variety of different crystal shapes are possible, but a specific substance generally takes the same geometric shape whenever it is crystallized.

Crystals can be grown from saturated solutions—solutions that have the maximum amount of solute dissolved in the solvent at a given temperature. The solvent is the substance which does the dissolving, and it is usually, but not always, the substance present in the greater amount. In this activity, water is the solvent (but it is not present in the greater amount). The solute is the substance (or substances) which is being dissolved—sugar in this activity.

The solubility in water of most, but not all, solids increases with temperature. Therefore, a saturated aqueous solution at the boiling point contains more solute than a saturated aqueous solution of the same substance at room temperature. In this activity, the water is heated so additional sugar can be dissolved. As the solution cools and the water evaporates over a period of days, the excess sugar comes out of the solution by forming crystals. Both the lowering of the temperature and the loss of solvent due to evaporation decrease the amount of solute which can remain in solution.

When the solute particles (in this case, sugar molecules) do come out of solution, they line up in an orderly pattern. The specific pattern depends on the nature of the solute. The number and size of the crystals depend on factors such as temperature, rate of solvent evaporation, and smoothness of the surface where crystallization occurs. The demonstration described in Extension 3 represents a simplified version of the process.

ASSESSMENT

Options:

- Have students draw a sequence of pictures representing the formation of sugar crystals from solution. Figure 2 shows possible starting and ending pictures.

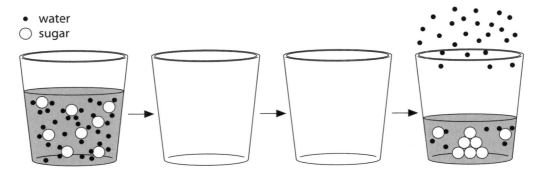

Figure 2: Have students draw a sequence of pictures representing the formation of sugar crystals.

- Ask how honey can be made to crystallize or if honey will crystallize faster in a capped jar or an uncapped jar.

- Tell the students that crystallized honey can be turned back to uncrystallized honey by heating it. Have them explain why this works.

CROSS-CURRICULAR INTEGRATION

Earth science:
- Students can examine geodes and read about how they are formed. (A geode is a hollow, usually spheroidal rock with crystals lining the inside wall.)
- Invite a geologist to speak to students about how crystals are formed in nature.
- If snow is available, have students use a magnifying lens to examine the ice crystals in snowflakes.

Social studies:
- Use Extensions 1 and 2 (making maple sugar and candy) as the basis for studying the importance of maple sap in pioneer cooking.

REFERENCE

"Crystal Growing" and "Crystal Gardens"; *Fun with Chemistry: A Guidebook of K–12 Activities;* Sarquis, M., Sarquis, J., Eds.; Institute for Chemical Education: Madison, WI, 1993; Vol. 2, pp 273–280.

CONTRIBUTORS

Charlotte Austin, Monroe Elementary School, Monroe, OH; Teaching Science with TOYS, 1993–94.

Peggy Kulczewski, Lincoln Elementary School, Monmouth, IL; Teaching Science with TOYS, 1994.

Cindy Waltershausen, Western Illinois University, Macomb, IL; Teaching Science with TOYS, 1994.

HANDOUT MASTERS

Masters for the following handouts are provided:
- Data Sheet
- Take-Home Activity

Copy as needed for classroom use.

Name _____   Date _____

Rock Candy Crystals
Data Sheet

Record your daily observations of the sugar solution with words or pictures.

Day 1	Day 2
Day 3	Day 4
Day 5	Day 6
Day 7	Day 8
Day 9	Day 10

Name _____   Date _____

Rock Candy Crystals
Take-Home Activity

Dear Adult Partner(s):

Your child has been growing and observing crystals in science. To assist your child in developing concepts about crystals and to share in the learning experience, please join your child looking at salt and sugar with a magnifying lens. Have your child draw what he or she observes in the spaces provided below. Then together write a short list of adjectives to describe each of the crystals.

Please send this completed Take-Home Activity back to school to let me know what you and your child did at home to study crystals.

Sincerely,

_____

picture of salt crystals	picture of sugar crystals
adjectives describing salt crystals	**adjectives describing sugar crystals**

Reproduced from *Investigating Solids, Liquids, and Gases with* **TOYS**, published by Terrific Science Press.

Crystals from Solutions

...Students use a variety of saturated solutions to grow crystals.

✔ **_Time Required_**

Setup	15	minutes
Performance	30–40	minutes (+ 1–2 days for crystals to grow)
Cleanup	5	minutes

✔ **_Key Science Topics_**

- capillary action
- crystals
- saturated solutions
- solids

Magic Tree kit

✔ **_National Science Education Standards_**

Science as Inquiry Standards:

- Abilities Necessary to Do Scientific Inquiry

  Students make systematic observations of crystal growing systems.

  Students decide how to collect and organize their data.

  Students base explanations of crystal-growing on their observations.

Physical Science Standards:

- Properties and Changes of Properties in Matter

  A characteristic property of crystalline substances is the specific, three-dimensional pattern of the crystal lattice.

✔ **_Additional Process Skills_**

• predicting	Students predict which type of cotton-tipped swab will be the best support for growing crystals.
• comparing/contrasting	Students compare and contrast results achieved with different crystal growing systems.

141

MATERIALS

For Getting Ready
Parts B, C, and D, per class
- Growing Solution ingredients:
  - 180 mL (12 tablespoons) laundry bluing (can be found in or ordered by most grocery stores)
  - 90 mL (6 tablespoons) table salt
  - 180 mL (12 tablespoons) water
  - 48 mL (3 tablespoons) household ammonia
- container(s) for Growing Solution
- spoon or stirring stick
- goggles

Part D, per group or class
- sandpaper
- (optional) food color
- scissors
- 3–4 cotton-tipped swabs with cardboard stems
- 3–4 cotton-tipped swabs with colored plastic stems

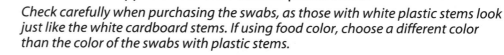

Check carefully when purchasing the swabs, as those with white plastic stems look just like the white cardboard stems. If using food color, choose a different color than the color of the swabs with plastic stems.

For Introducing the Activity
Per class
- magnifying lenses or microscope
- Magic Tree® or samples of various crystals, including salt and sugar

Magic Tree kits are available from Toysmith (#8309), 6250 S. 196th St., Kent, WA 98032; 800/356-0474.

For the Procedure
Per student
- goggles

Part B, per group or class
- 15–30 mL (1–2 tablespoons) Growing Solution (prepared in Getting Ready)
- tablespoon measure or 50-mL graduated cylinder
- blotter paper
- water-soluble ink pens
- scissors
- jar lid or Petri dish lid, or 150- to 250-mL beaker or similar-sized jar
- (optional) 9-ounce clear plastic cup and clear tape

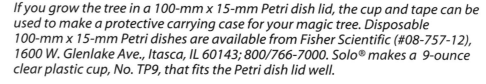

If you grow the tree in a 100-mm x 15-mm Petri dish lid, the cup and tape can be used to make a protective carrying case for your magic tree. Disposable 100-mm x 15-mm Petri dishes are available from Fisher Scientific (#08-757-12), 1600 W. Glenlake Ave., Itasca, IL 60143; 800/766-7000. Solo® makes a 9-ounce clear plastic cup, No. TP9, that fits the Petri dish lid well.

Part C, per group or class
- 30–45 mL (2–3 tablespoons) Growing Solution (prepared in Getting Ready)
- tablespoon measure or 50-mL graduated cylinder
- miniature pie pan or other shallow container
- charcoal briquet
- red, blue and green food colors

Part D, per group or class
- 15–30 mL (1–2 tablespoons) Growing Solution (prepared in Getting Ready)

Part D may require more solution depending on the size of the container.

- tablespoon measure or 50-mL graduated cylinder
- miniature pie pan, small clear plastic cup, or other shallow container
- oil- or plastic-based clay
- 6–8 swab halves with cardboard stems (prepared in Getting Ready)
- 6–8 swab halves with plastic stems (prepared in Getting Ready)

For the Variation
All materials required for the Procedure except
- use a porous surface for the crystal growing surface
- other variations as suggested by students

SAFETY AND DISPOSAL

Household ammonia and its vapor can damage the eyes. Eye protection is required even when using dilute solutions. Use only in a well-ventilated area. Should contact with eyes occur, rinse the affected area with water for 15 minutes. Medical attention should be sought while rinsing. The household ammonia used in this activity can be diluted with water and flushed down the drain. Unused solutions can be saved for further use.

Laundry bluing alone or in the Growing Solution will stain hands, clothing, and utensils. Use appropriate caution.

GETTING READY

Prepare the Growing Solution (per class, per part):

The amounts listed in Step 1 should make enough solution for a class to do one part of the Procedure as a hands-on activity. More solution may be required in each part depending on the size of the container.

1. While wearing goggles, place the following Growing Solution ingredients in a container and mix:

   - 180 mL (12 tablespoons) laundry bluing
   - 90 mL (6 tablespoons) table salt

- 180 mL (12 tablespoons) water
- 48 mL (3 tablespoons) household ammonia

2. Stir well to dissolve as much of the salt as possible.

3. Cover the solution so that it will not evaporate before use.

Prepare swab halves for Part D:

1. Roughen the surfaces of the swabs' stems with sandpaper.

2. Cut both the cardboard and plastic-stemmed swabs in half.

3. (optional) Soak the swab halves in food colors and allow them to dry.

INTRODUCING THE ACTIVITY

Options:

- Prepare a commercial Magic Tree by following the package directions.

- Show the students a variety of crystals. Give them the opportunity to observe several different crystals, such as sugar and salt, with a magnifying lens or microscope. Explain that they will be growing very tiny crystals that will not look as regular as, for example, salt, but more like powdered sugar.

PROCEDURE

You may wish to have each group do all parts of the Procedure, or you may select groups to do each part and then have groups compare results.

Part A: Systematic Observations

1. Discuss the importance of making systematic observations in a scientific investigation. As a class, brainstorm a list of questions students might try to answer for each crystal growing system as they observe the crystals forming. Ideas might include the following:
   - What changes occur right away?
   - How much time elapses before the first crystal is observed?
   - How many hours/days do crystals continue to grow?
   - How long before all the liquid in the Growing Solution has evaporated?
   - How large do the crystal formations get?

2. Discuss the importance of developing an organized method for making and recording observations. Have each group develop a plan for doing so.

Part B: Homemade Magic Tree

Make a homemade "magic tree" as follows:

1. Cut two tree shapes from blotter paper following the patterns shown in Figure 1.

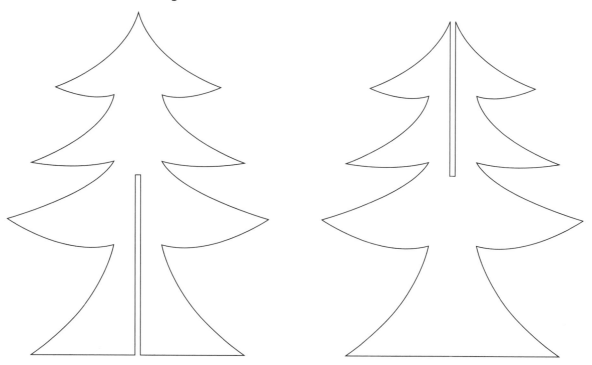

Figure 1: Cut two tree shapes out of blotting paper.

2. Use water-soluble marking pens to color the tips of the "tree branches."
This color will be picked up by the crystal solution and will color the crystals.

3. Slide the pieces together to make a tree.

4. Put 1–2 tablespoons of the Growing Solution prepared in Getting Ready in a puddle in a container and stand the tree in the puddle.
If you wish to preserve the tree as described in Step 6, use the larger half of a Petri dish as the container.

5. Place the tree in a location where it will not be disturbed. Let the tree system stand undisturbed for several hours.
It will take at least several hours for the crystals to grow. High humidity or cold room temperature may slow the evaporation rate and increase the time required to grow the crystals.

6. (optional) If you wish to preserve the tree, place a clear plastic cup upside down over the tree. Tape the cup securely to the Petri dish to make a protective carrying case for the tree.

Part C: Crystal Garden

Prepare a charcoal briquet crystal garden as follows:

1. Have each group place its charcoal briquet in a miniature pie pan or other shallow container.

2. Slowly pour 30–45 mL (2–3 tablespoons) of the Growing Solution over the briquet, wetting the entire surface.

3. Add drops of red, blue, and green food color to different regions of the briquet. A few isolated regions of color work better than mixing colors.

4. Place the charcoal briquet in a dry place where it will not be disturbed. It will take one or two days for the crystals to grow. Do not touch the crystals that develop since they are very powdery, fragile, and easily destroyed.

High humidity or cold room temperature may slow the evaporation rate and increase the time required to grow the crystals.

Part D: Crystal Flowers

1. Have each group flatten a piece of clay on the bottom of a miniature pie pan, cup, or other shallow container. (The clay should be thick enough to support the cotton-tipped swabs and should cover enough area to allow space between the swabs. See Figure 2.)

2. Show students the two varieties of cotton swabs: those with cardboard stems and those with plastic stems. Ask them to predict which will be best for growing crystals.

3. Give students the dry swabs and have them insert them into the clay. (See Figure 2.)

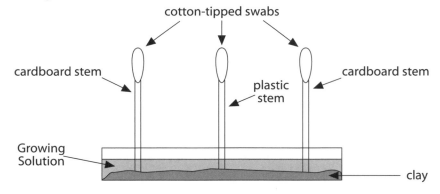

Figure 2: Insert swabs into the clay and add Growing Solution.

4. Add 1–2 tablespoons Growing Solution to each container. If the lower portion of each swab is not in the solution, add additional solution until they all are.

5. Place the containers in a location where they will not be disturbed and observe regularly until the crystal flowers have grown. (This may be a week or more.)

6. Ask the students which swabs worked better for growing the crystal flowers. *The swabs with the cardboard stems worked better.* Have them explain why this is the case. Discuss capillary action and wicking, defining the terms if the students are not familiar with them. (See the Explanation.)

Part E: Class Discussion

1. Discuss the similarities and differences between the different crystal growing systems. Bring out the idea that they all used the same Growing Solution and all had some kind of porous surface that enabled liquid to wick from the container (except the plastic-stemmed swabs). Give each group a chance to share their observations with the class. Have students propose explanations for the differences, citing specific observations as evidence for their ideas.

VARIATION

Try making changes in the Procedure, predicting the effects of the changes, and then testing. For example, compare the rate of crystal formation in a sunny window with the rate of formation in a cooler, darker area. What happens if water is left out of the Growing Solution? Try using a different porous object, such as a piece of sponge, as the crystal growing surface. Challenge your students to come up with other factors to test.

EXPLANATION

The following explanation is intended for the teacher's information. Modify the explanation for students as required.

A crystal is a solid that has a definite geometric shape due to the regular repeating pattern of the atoms, molecules, or ions that make up the substance. During crystallization, atoms, molecules, and/or ions align themselves in a specific, three-dimensional pattern called a crystal lattice; this lattice is a characteristic property of crystalline substances. The crystal lattice is responsible for the geometric patterns found in many crystals but is present even when obvious crystal shapes are not observed (as in this activity).

Crystals can be grown from saturated solutions, solutions that contain the maximum amount of solute (salts in this activity) that the solvent (water) can dissolve at a given temperature. The blotter paper, briquet, or cotton-tipped swabs provide a surface on which crystallization can occur.

The solution moves up the paper or cardboard or through the briquet to the surface by a type of capillary action called wicking. Wicking is not possible with the nonporous stems on the plastic cotton-tipped swabs. Capillary action occurs when a liquid moves through the pores of a substance. Wicking is a process similar to that occurring in candles, where the melted wax travels up through the pores of the wick as the candle burns. When the ammonia/bluing/salt solution reaches the surface of the paper tree or briquet, the water in the solution begins to evaporate. As the water evaporates, the salts in the solution are left behind as soft, flower-like crystals.

REFERENCES

"Crystal Gardens;" "Growing Crystals;" *Fun with Chemistry: A Guidebook of K–12 Activities;* Sarquis, M., Sarquis, J., Eds.; Institute for Chemical Education: Madison, WI, 1993; Vol. 2, pp 273–280.

Hillman, H. *Kitchen Science;* Houghton Mifflin: Boston, MA, 1989.

Sherwood, E.; Williams, R.; Rockwell, R. "Crystal Flower Garden;" *More Mudpies to Magnets: Science for Young Children;* Gryphon: Mt. Ranier, MD, 1990; p 39.

CONTRIBUTORS

Kari Dillman, Westlake Elementary School, New Carlisle, OH; Teaching Science with TOYS, 1995–96.

Pete Peterson, Shelby High School, Shelby, MI; Teaching Science with TOYS, 1996.

Gary Sundin, Grand Rapids Union High School, Grand Rapids, MI; Teaching Science with TOYS, 1996.

Changes of State

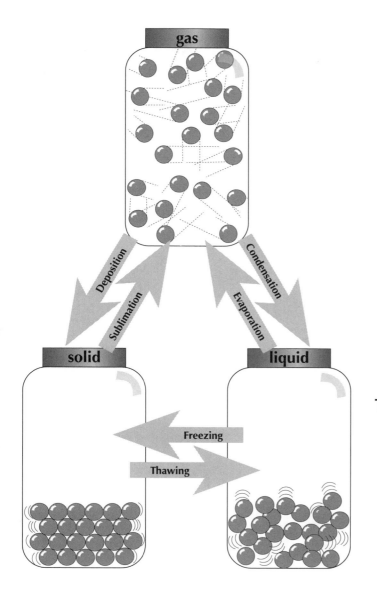

Crystals by Freezing

Boiling Water in a Paper Pot

Boiling Liquids in a Syringe

Boiling Water with Ice

Liquid to Gas In a Flick

Disappearing Air Freshener

A Cool Phase Change

Using Dry Ice to Inflate a Balloon

The Phase Changes of Carbon Dioxide

Balloon-into-a-Flask Challenge

Crushing an Aluminum Can

Hats Off to the Drinking Bird

✔ *Hands-On Activity*
Demonstration
Learning Center

Crystals by Freezing

...Students use the freezing process to form crystals of different sizes.

✔ Time Required

Setup	10	minutes
Performance	10–15	minutes for Part A
	10–15	minutes for 2 days for Part B
	30	minutes for Part C
Cleanup	10	minutes

✔ Key Science Topics

- crystals
- phase change
- solids

Smooth ice cream has very small ice crystals.

✔ National Science Education Standards

Science as Inquiry Standards:

- Abilities Necessary to Do Scientific Inquiry

  Students make systematic observations of frozen treats under different conditions.

  Students review observations and form a logical argument about the cause-and-effect relationships in the investigation.

Physical Science Standards:

- Properties and Changes of Properties in Matter

  A characteristic property of liquids is that they freeze when cooled below their freezing point.

- Transfer of Energy

  Heat moves in predictable ways, flowing from warmer objects to cooler ones.

✔ Additional Process Skills

- observing   Students observe the textures and tastes of various frozen treats.
- predicting   Students predict the texture of crystals in freezer pops and ice cream.

MATERIALS

For the Procedure, Part A
Per student
- 2 prepackaged freezer pops, one frozen and one non-frozen

> *These products are available under various names, including Pop Ice® and Fla-Vor-Ice®.*

For the Procedure, Part B
Per class
- 2 containers of ice cream
- ice cream scoop
- access to a freezer

Per student
- 2 small, disposable cups
- 2 disposable spoons

For the Procedure, Part C
Per student
- 2 small, disposable cups
- 2 disposable spoons

Per group
- 1 1-gallon zipper-type freezer bag
- 4 1-quart zipper-type freezer bags
- 1-cup measure
- set of measuring spoons
- 10 cups crushed ice

> *Bagged ice sold at grocery and convenience stores can be used if the pieces are crushed fairly small. If large pieces are used, the ice cream mixture may not freeze.*

- 1 cup rock salt
- 2 cups milk

> *Higher-fat milk will produce a thicker pudding pop taste; skim milk will produce more of an ice cream taste.*

- 2 tablespoons sugar (Omit sugar if chocolate milk is used.)
- ½ teaspoon vanilla
- serving spoon

Per class
- bucket for discarded water

SAFETY AND DISPOSAL

Because this activity involves eating, it should NOT be performed in the laboratory, nor should glassware or equipment be used that has previously been used for science experiments.

GETTING READY

Freeze one freezer pop for each student.

INTRODUCING THE ACTIVITY

Discuss the similarities and differences between frozen treats, such as Popsicles™ and ice cream. Discuss how both are made and have students bring favorite recipes to class if desired. *Both require a freezing process, and both result in the formation of ice crystals. Popsicles depend on the crystals growing very large, and ice cream depends on the crystals NOT being allowed to grow large.* Ask students if they've ever eaten ice cream that has thawed and then refrozen without constant mixing. Have them describe the result.

PROCEDURE

You may wish to have students complete the Data Sheet (provided) as the class performs the Procedure.

Part A: Freezer Pops

1. Without opening, have students examine a non-frozen commercial freezer pop. Read the contents and directions for freezing from the package. Have students predict the texture of the freezer pop after freezing.

2. Show students the frozen freezer pops and again examine them without opening them. Discuss the phase change that has occurred and if energy was added or removed.

3. Allow the frozen freezer pop to warm to room temperature and examine the contents. Discuss the phase change that occurred and the energy transferred.

Part B: Melting and Refreezing Ice Cream

1. Allow a container of ice cream to melt. Discuss the phase change that is occurring.

2. Pour the contents into small disposable cups so that each student will be able to have a taste in Step 4.

3. Place the cups into the freezer and refreeze the ice cream overnight. Do not stir or agitate the contents during the freezing process. Have students predict the texture of the refrozen, unstirred ice cream.

4. Once the ice cream has refrozen, try the following comparative taste test: Provide each student with a sample of regular ice cream and a sample of the refrozen ice cream. Have them discuss the similarities and differences.

Part C: Making Ice Cream in a Zipper-Type Bag

Have students make ice cream in a zipper-type plastic bag:

1. Measure 1 cup milk, 1 tablespoon sugar (omit if chocolate milk is used), and ¼ teaspoon vanilla into each of two 1-quart zipper-type freezer bags and seal the bags. Label one bag "unshaken" and the other "continuously shaken."

2. Place each bag in another 1-quart zipper-type bag to prevent leaks. Predict the texture of the ice cream in the unshaken bag and the ice cream in the continuously shaken bag.

3. Place the bag labeled "unshaken" in a freezer to freeze.

4. Each group should place the bag labeled "continuously shaken" into a 1-gallon zipper-type freezer bag, add 3 cups crushed ice and ½ cup rock salt, and seal the bag. Take turns shaking and kneading the second bag. Continue shaking the bag and replacing the crushed ice and salt until the ice cream is frozen. This process should take 5–10 minutes.

   When water accumulates in the 1-gallon bag, have students pour the water into the bucket and add more crushed ice and rock salt in the following ratio: 3 tablespoons salt to 1 cup crushed ice.

5. Wipe off any salty water from around the mouth of the shaken bag. Carefully spoon the ice cream into separate cups and try a comparative taste test. Compare the texture of the unshaken ice cream with that of the ice cream that was shaken.

6. While the students are eating, have them think back to the two samples tasted in Part B. Discuss the similarities and differences between the commercial ice cream, the melted and refrozen commercial ice cream, and their two versions of homemade ice cream. Be sure to discuss differences in the sizes of ice crystals in the samples.

EXPLANATION

> *The following explanation is intended for the teacher's information. Modify the explanation for students as required.*

The energy of a particle is related to its temperature; as the temperature increases, the energy increases. The temperature of a liquid is related to the average kinetic energy of the particles of the liquid.

When a substance is in the liquid state, the particles touch but can move relative to neighboring particles. This motion is responsible for the fluidity of liquids. Attractions between particles are strong enough to keep the particles in contact with each other but not so strong that the particles are locked into a certain position as they are in the solid state. When a liquid is cooled, the average energy of the particles decreases. At the freezing point, the attractive forces between particles overcome the energy responsible for the movement of the particles, and the particles no longer change position with respect to neighboring particles. At a pressure of 1 atmosphere, water freezes at 0°C.

The size of the ice crystals in ice cream depends on whether the mixture is stirred as it freezes. Milk is an emulsion of water and fat. Stirring or shaking the ice cream mixture causes the water in the milk to remain as tiny droplets in an emulsion with the fat. When these droplets freeze, the resulting ice crystals are tiny, giving the ice cream a smooth texture. If the ice cream mixture is not continually stirred or shaken as it freezes, some of the water separates from the fat particles in the milk and collects in larger drops, which freeze into larger ice crystals, giving the ice cream a coarser texture. When ice cream is melted and then refrozen without stirring, as in Part B, some water separates from the emulsion. The water refreezes into larger crystals, giving the ice cream a grainier texture than it had before.

Unlike ice cream, freezer pops contain no fat; the ingredients form a solution instead of an emulsion. The components of the solution do not separate when in the liquid state. As a result, the freezer pop can be frozen and refrozen many times and still retain the same texture.

In Part C of this activity, students start with a liquid ice cream mixture. Cooling the mixture causes the liquid to freeze into solid ice cream. In order for the mixture to cool, heat must be removed from it. Heat is transferred from a warmer substance to a colder substance when these substances are in contact with each other. The rate at which heat can be removed is increased by increasing the contact between the liquid ice cream mixture and the ice. A cold solution would be more efficient at transferring heat from the mixture than a cold solid would be, because the cold liquid could completely surround the ice cream mixture, putting more surface area of the mixture in contact with the cold substance. So, the mixture is cooled by placing it in contact with an ice and salt mixture.

Ice cream is a mixture, and the freezing point of a mixture is always lower than the freezing point of the pure solvent. This phenomenon is called freezing point depression. In this case, the solvent is water. Since the freezing point of the ice cream mixture is lower than the freezing point of water (0°C), an ice-water bath (which has a temperature of 0°C) is not cold enough. To freeze ice cream, the temperature must be lowered to about −10 to −15°C. How can you cool the ice cream mixture to this temperature? It cannot be cooled using ice alone. While the temperature of ice from a freezer is lower than 0°C, when it melts, the temperature of the ice-water bath that results is 0°C, which is not low enough to freeze the ice cream mixture.

In this activity, the ice cream mixture is frozen using an ice-salt mixture. As the ice melts, the resulting ice-salt mixture drops to a temperature of −10°C or lower. The salt keeps the salt-water mixture from freezing and also allows the salt-water mixture to remain a liquid at a temperature low enough to freeze the ice cream mixture. Other applications of freezing point depression include adding antifreeze to car radiators to keep the water liquid when the temperature drops below freezing in the winter and adding salt to icy roads to melt ice, even when the temperature is below 0°C.

CROSS-CURRICULAR INTEGRATION

Home, safety, and career:
- Ice crystals are an important component of ice cream and frozen foods. Have students conduct library research to discuss the role that crystal size plays in freezing foods.

Life science:
- Have students research the role of crystal size in determining how biological organisms withstand subfreezing temperatures.
- Discuss the Cincinnati Zoo's Center for Research of Endangered Wildlife (CREW). CREW scientists help preserve endangered species by freezing their sperm, eggs, and embryos and later thawing and implanting them into host mothers of related (but not endangered) species. A critical factor in the cryopreservation process is controlling the size of the ice crystals that form in the cells. General information related to the CREW project is available by calling 513/961-2739. Additionally, teachers within driving distance of Cincinnati, Ohio, can arrange a class visit to CREW.
- Discuss how frostbite damages living cells and how it can be prevented.

REFERENCES

Gibbon, D.L.; Kennedy, K. "The Thermodynamics of Home-Made Ice Cream," *Journal of Chemical Education.* 1992, *69* (8), 658–661.

Hillman, H. *Kitchen Science;* Houghton Mifflin: Boston, MA, 1989; p 116.

McGee, H. *On Food and Cooking: The Science and Lore of the Kitchen;* Scribner's: New York, 1984.

McMasters, M. "Let's Make Ice Cream!" *AIMS Newsletter,* January 1992; 18.

HANDOUT MASTER

A master for the following handout is provided:

- Data Sheet

Copy as needed for classroom use.

Name _____   Date _____

Crystals by Freezing
Data Sheet

Part A

1. Observe the unfrozen freezer pop. Predict the texture the freezer pop will have after freezing.

   _____

   _____

2. Observe the frozen freezer pop. Describe the texture of the freezer pop after freezing.

   _____

   _____

3. What phase change occurred? Was energy added to the freezer pop or removed from the freezer pop? Explain your answers. _____

   _____

   _____

Part B

1. Predict the texture of the refrozen, unstirred ice cream. _____

   _____

   _____

2. Observe the refrozen ice cream. Describe the texture of the ice cream after refreezing.

   _____

   _____

3. Compare the taste of regular ice cream with the taste of the refrozen ice cream. Discuss the similarities or differences. _____

   _____

   _____

Reproduced from *Investigating Solids, Liquids, and Gases with **TOYS***, published by Terrific Science Press.

Part C

1. Predict the textures of the unshaken ice cream and the continuously shaken ice cream. Will one be smoother than the other? _____

_____

_____

2. Observe the unshaken and continuously shaken ice creams. Describe the texture of each.

_____

_____

_____

3. Describe any similarities and differences in texture, crystal size, and taste among the three types of ice cream (refrozen, unshaken, and continuously shaken). _____

_____

_____

Putting It All Together

Write a paragraph comparing the various frozen treats you observed in this activity. Be sure to discuss how crystal size affects the taste and texture of the treat.

_____

_____

_____

_____

_____

_____

_____

_____

Boiling Water in a Paper Pot

...Students boil water in an unlikely container.

✔ Time Required

Setup   5 minutes
Performance   15–20 minutes
Cleanup   5 minutes

✔ Key Science Topics

- boiling
- endothermic changes
- kindling temperature
- phase changes

✔ Student Background

Students should be familiar with the process of boiling water.

✔ National Science Education Standards

Science as Inquiry Standards:

- Abilities Necessary to Do Scientific Inquiry

  Students develop an explanation based on their observations.

Physical Science Standards:

- Properties and Changes in Properties of Matter

  A characteristic property of water is a boiling point of 100°C at normal atmospheric pressure. The water temperature does not rise beyond 100°C until all the water has converted to steam.

- Transfer of Energy

  Heat transfer is predictable; it always flows from warmer objects to cooler ones.

Water boiling in a paper pot

✔ Additional Process Skills

• observing	Students observe as water is heated in an unlikely container.
• predicting	After watching an empty paper cup ignite, students predict what will happen when a paper cup is heated with water in it.

MATERIALS

For Introducing the Activity
Per class
- paper muffin cup or non-waxed paper cup
- container of water
- candle
- match
- tongs

For the Procedure
Part A, per class
- paper muffin cup or non-waxed paper cup
- ring stand and 5-cm (2-inch) diameter ring
- matches
- candle or Bunsen burner
- goggles

Part B, per class
- 2 balloons
- matches

Per student
- goggles

SAFETY AND DISPOSAL

Boiling water and steam can cause severe burns. Be sure that the water in the pot is cool before removing the pot. Stop heating the paper pot before all of the water has boiled away; without water present, the paper will char and ignite. No special disposal procedures are required.

GETTING READY

Assemble the ring stand and burner. The ring should be about 5 cm (2 inches) above the top of the flame.

INTRODUCING THE ACTIVITY

Have a container of water close by to extinguish the flame if needed.

Tear off part of the paper from a paper muffin cup or non-waxed paper cup. Light a candle. Ask the students to predict what will happen when you hold the paper in the flame of the lit candle. Using tongs, hold the paper in the flame to show that the paper burns.

Ask the students to speculate on whether adding water to the paper cup will cause different results when it is heated with an open flame. Use this question as a lead-in to the activity.

PROCEDURE

Part A: Boiling Water in a Paper Pot

1. Place the paper muffin cup or non-waxed paper cup into the ring. DO NOT light the heat source yet, but use the position of the heat source to adjust the height of the ring so that the tip of the flame will just touch the bottom of the cup. (See Figure 1.)

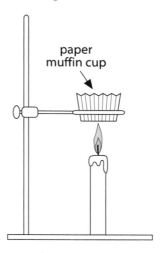

Figure 1: The setup for boiling water in a paper pot

2. Add water to the paper cup to a depth of about 1 cm (about ⅓ inch). *Have water ready to add to the pot if too much boils away. Water must be kept in the pot at all times; otherwise the paper will burn.*

3. Light the heat source and carefully slide it under the paper cup containing the water. Ask students to watch for evidence that the water is boiling.

Boiling water and steam can cause severe burns. Be careful not to hold your hand above the pot when the water is boiling or to handle the pot until the water cools.

4. Once the water begins to boil, turn off the flame and allow the water to cool.

5. (optional) Once cool, pour off the water and show the students the soot that has collected at the bottom of the cup. You can gently rub off most of the soot to show that the cup has not burned.

6. Through a class discussion, develop the idea that energy is transferred from the heat source to the water. Encourage students to base their comments on observations from Introducing the Activity and Steps 1–5 of Part A. Ask, "What would happen if all the water boiled

away? Why do you think so?" *The pot would catch fire because the water would no longer be absorbing the energy from the flame.* Discuss what would happen if you put an ice cube in the pot instead of liquid water. *The paper might catch on fire since the ice cube does not spread out in the cup.* Ask students to suggest other possible variables.

Part B: Heating Water in a Balloon

1. Light a match and use it to heat an inflated balloon with no water in it; the balloon will burst almost immediately once the flame contacts the rubber. Ask students to describe what they observed.

2. Place a small amount of water in a balloon, inflate the balloon, and tie it off. Have students predict what will happen if a balloon with water in it is heated.

3. Light a match and use it to heat the balloon directly below the water. (See Figure 2.)

 Do not hold the balloon in the flame for too long; the balloon may burst.

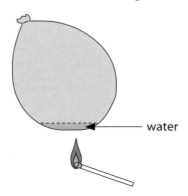
water

Figure 2: Heat the balloon directly below the water.

4. Discuss why the results are different when the balloon is heated with water in it and without water in it.

 EXPLANATION

The following explanation is intended for the teacher's information. Modify the explanation for students as required.

Although a candle (or a burner) provides sufficient heat to burn paper, the paper pot does not burn because of the liquid water that remains in it. The liquid water absorbs the heat and keeps the temperature of the paper below its kindling temperature, around 230°C.

As the water is heated, it increases in temperature. When the temperature of the water reaches 100°C, it begins to boil. Boiling is an endothermic process, which means that energy is absorbed from the surroundings. At the boiling point, all the absorbed energy is used to change the water from a liquid to a gas, so its temperature remains constant at 100°C until all the water has been converted to steam. As long as liquid water remains, the temperature of the paper pot will not exceed 100°C, and the paper will remain well below its kindling temperature.

If all the water boils away, the temperature of the paper will increase, since there is no water remaining to absorb the heat being added from the candle. Once the water is gone, the paper pot quickly chars and ignites.

When a match is used to heat a balloon filled with air, the rubber softens and the balloon bursts almost as soon as the flame comes in contact with the rubber. When water is in the balloon, the heat from the match is absorbed by the water, which keeps the rubber from softening so much that it bursts.

ASSESSMENT

Have students write a brief discussion comparing the balloon system to the paper cup system, including a complete science explanation as part of the discussion.

CROSS-CURRICULAR INTEGRATION

Language arts:

- Have students write a story in which a character boils water using a container that would ordinarily burn or melt when heated. Instruct students to work the science explanation into their story.
- Suggest that students read one of the following books and relate some aspect of the story to the science activity.
  - *Hatchet,* by Gary Paulsen (Bradbury, ISBN 0-14-032724-x)
    After a plane crash, 13-year-old Brian spends 54 days in the wilderness learning to survive with only the aid of a hatchet, the clothes on his back, and the things in his pockets—which include a piece of paper.
  - *Fahrenheit 451,* by Ray Bradbury (Ballantine, ISBN 0-345-34296-8)
    Fahrenheit 451 is the temperature at which book paper catches fire and burns. Montag, a regimented fireman in charge of burning forbidden volumes, meets a revolutionary schoolteacher who dares to read. Suddenly Montag finds himself a hunted fugitive, forced to choose not only between two women, but also between personal safety and intellectual freedom.

REFERENCES

"Boiling Water in a Paper Pot," *Fun with Chemistry: A Guidebook of K–12 Activities;* Sarquis, M., Sarquis, J., Eds.; Institute for Chemical Education: Madison, WI, 1991; Vol. 1, pp 173–176.

Shakhashiri, B.Z., *Chemical Demonstrations;* University of Wisconsin: Madison, WI, 1989; Vol. 3, pp 239–241.

Boiling Liquids in a Syringe

... Students observe that water and other liquids will boil at temperatures below their normal boiling point if the pressure is reduced.

✔ Time Required

Setup   15–20 minutes
Performance 10–15 minutes*
Cleanup   5 minutes
*Additional time is required if done as a hands-on activity.

✔ Key Science Topics

- boiling
- phase changes
- relationship of pressure to boiling
- vapor pressure

✔ Student Background

This activity is most effective if students have a basic understanding of normal boiling points and the boiling process. It can also be used to introduce or reinforce the concept that boiling temperature is dependent on pressure.

Liquid boiling in a syringe

✔ National Science Education Standards

Science as Inquiry Standards:

- Abilities Necessary to Do Scientific Inquiry

  Students identify questions that can be answered through scientific investigations.

  Students design and conduct a scientific investigation, identifying and controlling variables and making accurate measurements.

Physical Science Standards:

- Properties and Changes in Properties of Matter

  A substance has a characteristic boiling point that is affected by the atmospheric pressure; for example, the boiling point changes with changes in pressure.

✔ Additional Process Skills

- observing   Students observe as water boils at a temperature below the normal boiling point.

- predicting   Students predict what will happen to the water in the sealed syringe when the plunger is pulled back.

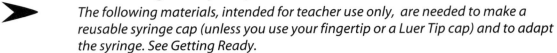

MATERIALS

For Getting Ready only

The following materials, intended for teacher use only, are needed to make a reusable syringe cap (unless you use your fingertip or a Luer Tip cap) and to adapt the syringe. See Getting Ready.

*   disposable syringe needle

An appropriate needle is available from Fisher Scientific (#14-826-5B), 1600 W. Glenlake Ave., Itasca, IL 60143; 800/766-7000. Many others will also work.

*   candle or Bunsen burner
*   2 pairs of forceps, tweezers, or pliers
*   small container of water
*   clear plastic 50- or 60-mL syringe

Syringes with a capacity between 50–60 mL are available from Fisher Scientific (60 mL #14-823-2D); they may also be available from a veterinarian or farm supply store. If you use a syringe with a Luer-Lok tip, you can make a reusable syringe cap as described in Getting Ready.

*   (optional) nail

For Introducing the Activity
Per class
*   beaker of water

For the Procedure
Part A, per class or group
*   syringe prepared in Getting Ready
*   hot tap water at least 40°C (104°F) in a cup or other container
*   room-temperature water in a clear plastic container large enough to submerge both hands in
*   paper towels
*   1 of the following unless you use your finger to seal the syringe (see Getting Ready):
    *   syringe cap prepared in Getting Ready
    *   Luer Tip cap

Luer Tip caps are available from Fisher Scientific (#14-826-76).

Part B, per student or group
*   syringe prepared in Getting Ready
*   1 of the following unless students use a finger to seal the syringe (see Getting Ready):
    *   syringe cap prepared in Getting Ready
    *   Luer Tip cap
*   materials listed by students for their experimental designs

- (optional) 10–15 mL of any or all of the following liquids:
  - ethyl alcohol (ethanol)
  - 70% ethyl rubbing alcohol
  - 91–99% isopropyl alcohol (isopropanol)
  - rubbing alcohol (70% isopropyl alcohol solution)
  - acetone
  - fingernail polish remover containing acetone
- (optional) cup
- (optional) tub of hot water

SAFETY AND DISPOSAL

If you choose to make a reusable syringe cap from a disposable needle, prepare the syringe cap (see Getting Ready) outside of class. Take care not to get melted plastic on your skin. Proper fire safety should be exercised, such as working on a fire-resistant surface and removing unnecessary flammable materials from the area. Long-haired people should tie hair back when working near a flame. Dispose of the needle by placing it in the plastic needle cover and wrapping tape over the open end. Place the wrapped needle cover in the trash.

The liquids used in Part B of the Procedure are flammable, and their vapors are irritating to the eyes and respiratory system. Use these liquids only in a well-ventilated area and keep flames away. The liquids are harmful if ingested, and they can cause severe damage to the eyes. Should contact with the eyes occur, rinse with water for 15 minutes and seek immediate medical attention.

Certain liquids are inappropriate for use in the activity. For example, methanol and duplicating fluid are toxic when absorbed through the skin or when the fumes are inhaled. For this reason, use only the liquids suggested in Materials. Liquids can be saved for future use or flushed down the drain with water.

GETTING READY

1. The activity requires that the needle end of the syringe be temporarily sealed by one of the following methods:

   a. Use a snugly fitting Luer Tip cap to seal the tip of the syringe. Slip the narrow part of the Luer Tip cap over the open tip of the syringe using pressure and a twisting motion. (See Figure 1.)

   Luer Tip caps can fly off the end of the syringe and shoot across the room if ample pressure is applied to a volume of air trapped in the syringe. Keep your finger on the Luer Tip cap when pushing the syringe barrel.

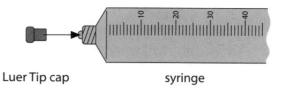

Luer Tip cap   syringe

Figure 1: Using a Luer Tip cap to seal the end of the syringe

 Take care not to get melted plastic on your skin.

    b. Prepare an inexpensive cap for each syringe as follows: Hold a disposable syringe needle with a pair of pliers, forceps, or tweezers, and use a candle or Bunsen burner to heat the needle close to where it enters the plastic connector. As the plastic begins to melt, pull the needle out with the second pair of pliers. (See Figure 2.) Drop the hot needle into the small container of water to cool. After the plastic has cooled, check the cap for leaks as follows: Place the cap on the syringe with the plunger completely in the "in" position. Pull the barrel until it is about halfway out and then release the barrel. If it returns to its original position, the syringe cap is properly sealed. If the barrel does not return to its original position, some air is in the syringe, and the cap leaks. Try heating the plastic tip again to get it to seal, or discard the cap and try again.

 If the plastic catches fire, extinguish the fire by dipping the cap in water. Once prepared, the cap may be saved and used again. Dispose of the needle as specified in Safety and Disposal.

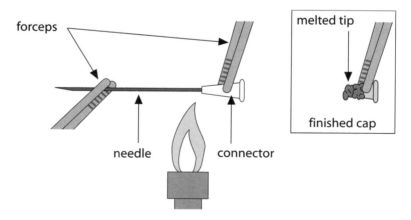

Figure 2: Remove the syringe needle to make a cap.

    c. During the activity, seal the syringe with your finger. This technique is simple, but it is somewhat awkward and is hard on the finger.

2. Because it is difficult to hold the plunger in the "out" position, you may want to adapt the plastic syringe so that the plunger can be held out automatically.

a. Adjust the plunger in the syringe barrel to read 50 mL (or the last calibrated point on the syringe), and mark the point on the plunger where it meets the barrel. Remove the plunger from the barrel.

b. Hold a nail with tongs or pliers, heat it with a Bunsen burner until it is hot enough to melt the plastic, and push it through the plunger at the marked point. Remove the nail. Now this nail or one with a slightly smaller diameter can be inserted through the hole to keep the plunger in the "out" position. (See Figure 3.)

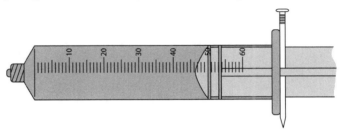

Figure 3: Use a nail to hold the plunger in the "out" position.

INTRODUCING THE ACTIVITY

1. Ask, "How do you know if a liquid is boiling?" *Bubbles form from within the liquid.*

Some of the first bubbles may be due to dissolved oxygen and nitrogen (air). When the temperature of a liquid is increased, the solubility of any dissolved gases decreases. However, once the liquid reaches its boiling point, the bubbles forming are bubbles of the gas phase of the liquid.

2. Show students a beaker of water and ask them what they would have to do to cause the water to boil. The most likely response will be, "Heat it to its boiling point." Ask the students what the boiling point of water is. They will probably answer, "100°C" (212°F), the boiling point of water at 1 atmosphere pressure. Ask students if the boiling point of water is always the same. Discuss the idea that boiling point varies with atmospheric pressure and thus altitude. For example, Denver is 1,600 meters (1 mile) above sea level where the atmospheric pressure is about 0.84 atmospheres, and in Denver, water boils at about 95°C (203°F).

3. Through discussion, bring out the idea that pressure and boiling point are related. Tell students that they will be exploring this relationship in the activity.

PROCEDURE

Part A: Demonstrating the Relationship Between Pressure and Boiling Point

1. Show students the syringe and ask how it might be useful for exploring the relationship between pressure and boiling point. (If students have

done the "Properties of Matter" or "Marshmallow in a Syringe" activities, remind them of these.) Bring out the idea that by pulling back the plunger of a sealed syringe you are lowering the pressure inside.

2. Demonstrate how to use the syringe and, if appropriate, how the Luer Tip or homemade cap seals the syringe and how the nail will hold the plunger out.

3. Pour hot tap water (at least 40°C) into a cup or other container. Tell students the temperature of the water.

4. With the tip open, push the plunger all the way into the syringe barrel. Put the tip end of the syringe into the water and pull back the plunger to draw a little over 10 mL water into the syringe.

5. Hold the syringe vertically, tip upward. Tap the syringe barrel to dislodge any air bubbles from the sides of the barrel and end of the plunger. Carefully push the plunger in until all of the air is expelled from the tip of the syringe.

➤ *If large air bubbles are still present, empty the syringe and repeat Steps 4 and 5.*

6. Tightly seal the end of the syringe with a Luer Tip cap, homemade cap, or finger. Dry the syringe with paper towels. (If the syringe is wet, you may lose your grip during Step 8.)

7. Ask students to predict what will happen if you pull the plunger back while the syringe is sealed.

8. Pull the plunger back to about the 50-mL mark, and, if desired, push the nail through the hole to hold the plunger out.

➤ *You may need to practice this step, as pulling the plunger out can be difficult. You may need to determine the hand position that works best for you. Have an assistant push the nail through the hole.*

9. Bubbles of water vapor will begin to form in the liquid. Though this bubbling is not violent, the medium-sized bubbles should be visible to those nearby. The students may think that the bubbles are caused by air leaking around the plunger. To verify that air does not leak into the syringe, remove the Luer Tip cap, homemade cap, or finger, and push the barrel in completely to empty the syringe. Seal the end again and put the entire syringe into a clear container of room-temperature water (hand and all). Pull out the plunger in a smooth, continuous motion, and hold it out for a minute. Remove the syringe from the water and show the students that no water has been pulled into the syringe.

➤ *If you have modified the syringe so the plunger can be held out with a nail, you can pull out the plunger as described and insert the nail.*

Part B: Exploring the Relationship Between Pressure and Boiling Point

1. Discuss observations from Part A. Remind students that the temperature of water used in Part A was about 40°C (104°F), far lower than the normal boiling point of water. Ask students what conclusion they could draw about the relationship between pressure and boiling point based on the results of the demonstration. Make sure that students grasp the idea that a liquid's boiling point decreases as the pressure above the liquid decreases.

2. Have students brainstorm questions about boiling that they can test themselves using the syringe. Questions could include the following:
   - Could we get cooler water to boil? How cold could the water be and still boil with this method?
   - Can we make other liquids boil? How do their boiling points compare to that of water?

3. Discuss experimental design as a class. Emphasize the importance of controlling variables. Establish what the variables would be for each question proposed, and discuss how students would control these variables. Also, develop a class definition of "boiling" for this investigation (e.g., the appearance of one or two bubbles, some bubbles rising to the liquid's surface, or many bubbles rising to the liquid's surface).

Do not allow students to conduct experiments that involve hazardous materials or procedures. If students choose to test liquids other than water, do not allow them to use liquids other than those listed in Materials. If you wish to warm the liquids for a more dramatic and quicker result, fill a tub with hot tap water and hold a cup of the liquid in the hot tap water for a few minutes. (Ethanol, pure acetone, or 91–99% isopropyl alcohol will boil more vigorously than 70% ethyl rubbing alcohol, 70% rubbing alcohol, or fingernail polish remover.) The liquids listed for this part of the activity are flammable. DO NOT warm them directly with an open flame or a hot plate. Use a hot tap water bath only in an area with good ventilation.

4. Have students design experiments they could do to test one of the proposed questions. Have them describe their designs on the Experimental Design and Data Sheet (provided), including a list of materials they will need.

5. Have students gather the necessary materials and perform their experiments, recording their results on the Experimental Design and Data Sheet. Have students draw conclusions to the questions they addressed, based on their data.

6. Have students present their results and conclusions to the class. If results varied widely between trials, discuss possible reasons for the variations and ways to achieve greater consistency. Discuss whether the conclusions students have drawn are valid, based on the experiments they have done.

EXPLANATION

> *The following explanation is intended for the teacher's information. Modify the explanation for students as required.*

Boiling involves a phase change from liquid to gas. At temperatures less than the boiling point, only molecules at the surface with sufficient energy to escape the attractions between molecules can escape the liquid and change to the gas phase. The gas phase particles exert a pressure which is called the vapor pressure. As the temperature increases and molecules become more energetic, more escape and cause an increase in the vapor pressure. A liquid is considered to be boiling when bubbles of its vapor can come from anywhere in the liquid. This occurs when the vapor pressure equals the external pressure. The normal boiling point of a liquid is the temperature at which it boils when the external pressure is equal to 1 atmosphere, 760 torr (mean air pressure at sea level). The normal boiling point of water is 100°C (212°F). If the external pressure is less than 1 atmosphere, as it is at high altitudes and in this activity when the syringe plunger is pulled to the extended position, the boiling point of water is lower than 100°C. For example, the boiling point of water in Denver, Colorado (which is 1.6 kilometers or 1 mile above sea level) is about 95°C (203°F).

If the external pressure is greater than 1 atmosphere, the boiling point of water is greater than 100°C. For example, pressure cookers speed up the cooking process by increasing the pressure to more than 1 atmosphere. The increase in pressure causes water to boil at a higher-than-normal temperature (above 100°C), and thus shortens the time needed to cook the food.

Table 1: Boiling Points of Water at Various External Pressures		
Pressure		Boiling Point
atmospheres	torr	° C
0.006	4.6	0
0.023	17.5	20
0.042	31.8	30
0.073	55.3	40
0.197	149.4	60
0.467	355.1	80
1.000	760.0	100
1.959	1489.1	120

The boiling point of water at various pressures is shown in Table 1. Other liquids exhibit similar relationships between external pressure and boiling point. The normal boiling points of the pure liquids used in this activity are as follows:

ethyl alcohol (ethanol)	78.5°C
isopropyl alcohol (isopropanol)	82.4°C
acetone	56.2°C

ASSESSMENT

• Provide students with a copy of Table 1 and have them write an answer to the following question: Could you boil an egg (so that it was free from salmonella) at the top of Mt. Everest? Instruct them to use observations from the activity and information from Table 1 as evidence for the answer. Provide students with the following information: an internal yolk temperature of about 82°C (180°F) is required to kill salmonella bacteria, which are commonly present in chicken eggs. The top of Mt. Everest is 8,848 m (29,028 feet) above sea level, and the pressure is about 0.3 atm. *The boiling point of water at the top of Mt. Everest is approximately 69°C (156°F), so the water would not get hot enough to heat the yolk to 82°C (180°F) and kill the salmonella before boiling off.*

CROSS–CURRICULAR INTEGRATION

Home, safety, and career:
• Discuss how pressure cookers shorten the time needed to cook food.

REFERENCES

"Boiling Liquids in a Syringe"; *Fun with Chemistry: A Guidebook of K–12 Activities;* Sarquis, M., Sarquis, J., Eds.; Institute for Chemical Education: Madison, WI, 1993; Vol. 2, pp 189–194.

McGee, H. *On Food and Cooking: The Science and Lore of the Kitchen;* Macmillan: New York, 1984; pp 67–68.

Shakhashiri, B.Z. *Chemical Demonstrations;* University of Wisconsin: Madison, 1985; Vol. 2, pp 81–84, 116.

HANDOUT MASTER

A master for the following handout is provided:
• Experimental Design and Data Sheet
Copy as needed for classroom use.

Name _____   Date _____

Boiling Liquids in a Syringe
Experimental Design and Data Sheet

1. Question _____

   _____

2. Design of the Experiment

   _____

   _____

   _____

   _____

   _____

   _____

   _____

3. Materials Required

   _____

   _____

   _____

   _____

4. Picture of the setup

5. Prediction

_____

_____

_____

_____

_____

6. Data

_____

_____

_____

_____

_____

7. Conclusions

_____

_____

_____

_____

_____

_____

_____

_____

_____

Hands-On Activity

✔ **Demonstration**

Learning Center

Boiling Water with Ice

...Students observe water boiling while it is being cooled by ice.

✔ **Time Required**

Setup	15	minutes
Performance	10	minutes
Cleanup	5	minutes

✔ **Key Science Topics**

- boiling
- relationship of boiling point to pressure

✔ **Student Background**

The activity is most effective if students have a basic understanding of the boiling process, including the relationship of boiling temperature to the pressure.

Cooling the flask with ice

✔ **National Science Education Standards**

Science as Inquiry Standards:

- Abilities Necessary to Do Scientific Inquiry

  Students think logically to distinguish between evidence and explanations.

Physical Science Standards:

- Properties and Changes in Properties of Matter

  Water has a characteristic boiling point at sea level (1 atmosphere pressure). The boiling point of water changes as the pressure changes.

✔ **Additional Process Skill**

- observing
      Students make observations of the water in a flask while the flask is being cooled.

MATERIALS

For Introducing the Activity
- hot plate
- pan of water
- thermometer

For the Procedure
Per class
- 500-mL to 1-L thick-walled, round-bottomed flask
- 1-hole rubber stopper to fit the above flask
- alcohol thermometer (A V-back thermometer will not work.)
- ring stand
- iron ring
- wire gauze
- clamp
- zipper-type plastic bag filled with ice water
- Bunsen burner
- heat-resistant gloves
- towel
- 2–3 drops of soap, glycerin, or vegetable oil

Per student
- goggles

For the Variation
- aspirator
- 100-mL thick-walled, round-bottomed flask
- 2-hole rubber stopper to fit flask
- glass tubing
- alcohol thermometer
- ring stand
- iron ring
- clamp
- heavy-walled vacuum tubing to connect aspirator to flask

SAFETY AND DISPOSAL

A thick-walled, round-bottomed flask MUST be used. A Florence flask or an Erlenmeyer flask may crack or even implode because of the reduced pressure inside the flask. DO NOT use a test tube.

Avoid using mercury thermometers because of the potential for breakage and the toxic nature of mercury. Take care when inserting the thermometer into the stopper. Use a towel to protect your hand. When inserting the thermometer into the rubber stopper, carefully follow the instructions in Getting Ready. The glass thermometer can break and cut your hand if insertion is not done carefully.

GETTING READY

The thermometer can break and cut your hand if you do not insert it carefully. Hold the thermometer as close to the stopper as possible to lessen the possibility of breaking the thermometer.

1. Lubricate the bulb end of the thermometer with soap, vegetable oil, or glycerin. Wrap the thermometer with a towel and carefully insert the thermometer with a twisting motion into the hole of the rubber stopper. (See Figure 1.)

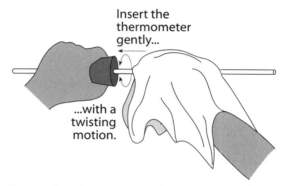

Insert the thermometer gently...

...with a twisting motion.

Figure 1: Wrap the thermometer in a towel and, grasping the thermometer as close to the stopper as possible, use a twisting motion to insert it into the hole of the rubber stopper.

2. Fill the flask about ⅓ full of water.

3. Insert the stopper into the flask, making sure that the thermometer does not touch the bottom of the flask.

4. Adjust the thermometer so that when you turn the flask upside down, the entire bulb is under water. (See Figure 2.)

Figure 2: Invert the flask to position the thermometer. Put the flask back in an upright position and remove the thermometer and stopper.

5. Set up the equipment as shown in Figure 3, but do not place the thermometer and stopper in the flask yet.

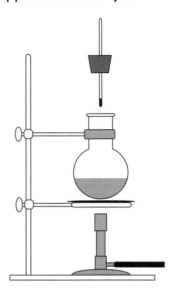

Figure 3: Set up the equipment.

INTRODUCING THE ACTIVITY

Show students a pan of water and ask them what they would have to do to cause the water to boil. The most likely response will be to heat it to its boiling point. Ask the students what the boiling point of water is. *Students will probably answer 100°C.* Place the pan on the hot plate, heat the water to the boiling point, and measure the temperature. Have a student read the thermometer to confirm that the water boiled at about 100°C. Ask students if they think the boiling point of water is always the same. Give students an opportunity to share ideas, but do not discuss them further at this time. Tell students they will be observing a demonstration that will provide some answers to that question.

PROCEDURE

DO NOT HEAT A STOPPERED FLASK; the stopper-thermometer assembly may be ejected or an explosion may result.

1. **With the stopper removed from the flask,** heat the flask with the Bunsen burner until the water is boiling vigorously. Ask students to describe what they see.

2. Turn off the Bunsen burner and wait until bubbles stop forming (about 30 seconds), then restopper immediately.

3. Insert the stopper-thermometer assembly into the flask and seal tightly.

4. Using heat-resistant gloves, turn the flask upside down and clamp it to the ring stand.

DO NOT COOL BELOW 70°C in the next step; the flask may implode if the pressure inside is lowered too far.

5. Place the zipper-type plastic bag of ice water on the top of the flask. (See Figure 4.) Ask, "What will this ice do to the water vapor inside the flask?" *Cool it.* "What happens to water vapor when it cools?" *It condenses (turns back into liquid).*

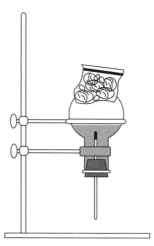

Figure 4: Place the bag of ice water on top of the inverted flask.

6. Tell students to watch closely and tell you when the water begins to boil. Read the temperature at which the water in the flask boils. (The temperature will be below 100°C.)

7. Once you have completed the demonstration, turn the flask right side up.

8. Carefully remove the stopper (it will be hard to remove) and listen for the sound of air rushing back into the flask.

9. Through a class discussion, lead students to understand why the water boiled at less than 100°C. (See the Explanation.) Discuss what evidence the demonstration provided for this explanation. Incorporate student observations and ideas into the discussion as much as possible.

10. Explain that what we usually consider the "normal" boiling point of water is the boiling point of water at 1 atm pressure (the approximate "mean" value at sea level). Tell students that Denver, Colorado, is 1,600 meters (1 mile) above sea level. Discuss whether water would boil at 100°C in Denver. Have students provide support for their ideas based on the demonstration they just observed.

VARIATION

If you have an aspirator, partially evacuate the flask to show students a different way of reducing pressure.

1. Set up the apparatus as shown in Figure 5. Be sure to use a round-bottomed, thick-walled flask and heavy-walled vacuum tubing.

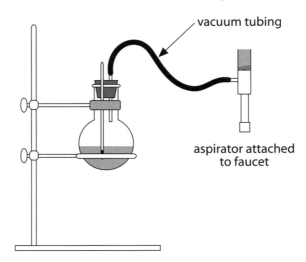

Figure 5: Attach an aspirator to the setup using heavy-walled vacuum tubing.

2. Remove the stopper from the flask and fill the flask ⅓ full of water.

3. Heat the water to 50–60°C, then turn off the burner.

4. Insert the stopper into the flask so that the thermometer bulb is in the water.

5. Turn on the aspirator.

6. Observe the temperature when the water begins to boil, and turn off the aspirator.

EXPLANATION

The following explanation is intended for the teacher's information. Modify the explanation for students as required.

Boiling involves a phase change from liquid to gas. At temperatures less than the boiling point, only molecules at the surface with sufficient energy to overcome the attractions between molecules can escape the liquid and change to the gas phase. The gas phase particles exert a pressure which is called the vapor pressure. As the temperature increases and molecules become more energetic, more escape and cause an increase in the vapor pressure. A liquid is said to be boiling when bubbles of its vapor can come from anywhere in the liquid. This occurs when the vapor pressure equals the external pressure. The normal boiling point of water, 100°C (212°F), is the boiling point at 1 atm pressure.

In this activity, the water is heated at atmospheric pressure until it is boiling. As the water boils in the flask, the steam (water vapor), pushes most of the air out of the flask. Stoppering the flask prevents air from re-entering. As the steam cools, it condenses back into liquid water. Using the bag of ice water increases the rate of condensation. A partial vacuum is created inside the flask, because the liquid takes up less space than steam. This partial vacuum exerts less pressure than was originally present. Since the pressure inside the flask is lower than atmospheric pressure, the water will boil at a temperature of less than 100°C.

The Variation uses an aspirator to reduce the pressure. At altitudes above sea level, the atmospheric pressure is less, so water boils at a lower temperature. Thus, the high-altitude boiling water has less energy to transfer to food, so food requires a longer cooking time at very high altitudes.

ASSESSMENT

Tell students that in pressure cookers the boiling point of water is higher than 100°C. Based on their understanding of the "Boiling Water with Ice" demonstration, have them explain how a pressure cooker works.

REFERENCES

"Boiling Water with Ice"; *Fun with Chemistry: A Guidebook of K–12 Activities;* Sarquis, M., Sarquis, J., Eds.; Institute for Chemical Education: Madison, WI, 1991; Vol. 1, pp 235–240.

Shakhashiri, B.Z. *Chemical Demonstrations;* University of Wisconsin: Madison, WI, 1985; Vol. 2, pp 81–84.

Hands-On Activity

✔ *Demonstration*

Learning Center

Liquid to Gas in a Flick

...Is butane a liquid or a gas at room temperature? This activity shows the butane's phase change from a liquid to a gas due to changing pressure.

✔ **Time Required**

Setup	5 minutes
Performance	20 minutes
Cleanup	5 minutes

✔ **Key Science Topics**

- phase change
- pressure
- temperature

✔ **Student Background**

This demonstration is most effective after students have discussed the three states of matter. The demonstration challenges them to question the factors that determine the state in which matter is found. It may also be used to stimulate thinking prior to discussion of the gas laws.

Butane lighter
with a transparent case

✔ **National Science Education Standards**

Science as Inquiry Standards:

- Abilities Necessary to Do Scientific Inquiry

  Students use appropriate tools and techniques to gather, analyze, and interpret data relating temperature to the rate of bubble release from the butane lighter.

  Students use mathematics to organize and present data from the different temperature trials.

Physical Science Standards:

- Properties and Changes in Properties of Matter

  A characteristic property of matter is that the state (gas, liquid, or solid) is affected by both pressure and temperature.

MATERIALS

For Introducing the Activity
Per class
- (optional) molecular motion simulator made from the following:
  ○ plastic "clamshell"-shaped food container (available at salad bars) or 2 clear plastic ready-made pie crust covers
  ○ tape
  ○ box of plastic beads

For the Procedure
Per class
- 2 metal cooking or alcohol thermometers
- ice
- butane lighter with a transparent case
- 3 clear plastic containers (large enough to immerse the butane lighter in; at least 1 that can be heated)
- overhead projector
- timing device with a second hand
- (optional) 6-inch (15-cm) test tube, transparent pill vial, and 50-mL graduated cylinder
- (optional) food color

For Variations and Extensions
❸ Per class
- an aerosol product that uses butane (or another hydrocarbon) as a propellant

SAFETY AND DISPOSAL

Butane gas is flammable and should not be used around an open flame. Do not allow younger students to use the butane lighter by themselves. Use a screwdriver to push apart the metal tabs in which the flint is mounted and remove the lighter wheel and the flint to prevent ignition of the gas by the lighter itself. Monitor the lighters carefully to prevent igniting or sniffing abuse.

Because of potential breakage of thermometers, metal cooking or alcohol thermometers should be used.

GETTING READY

1. Fill the clear plastic container with room-temperature water for Part A, making sure the water will not overflow the container when immersing your hand.

2. (optional) Add a few drops of food color to the water.

3. (optional) If you desire to simulate molecular motion in Introducing the

Activity, prepare the simulator as follows: pour enough plastic beads into the plastic clamshell-shaped container or one of the pie crust covers so that there is a single layer of plastic beads that covers ⅓–½ of the bottom of the container. (See Figure 1.) If the pie crust cover is used, put a second cover on top. If the clamshell-shaped container is used, close the top. Tape the two halves or the bottom and top together.

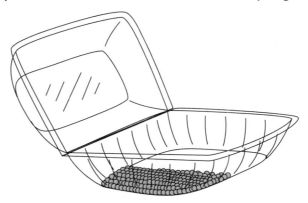

Figure 1: Prepare a molecular motion simulator.

4. Prepare the warm-water bath for Part B by mixing hot and cold tap water to get a temperature of 40–45°C or by heating water to a temperature of 40–45°C. Check the temperature of the water just prior to doing Part B.

 The bubbles form too quickly to count if the temperature is greater than about 45°C.

5. Prepare the cold-water bath for Part B by placing ice and water in a container. The temperature should be about 10°C.

 The container of water used in Part A can serve as the room-temperature bath in Part B.

INTRODUCING THE ACTIVITY

Use the overhead projector and the setup prepared in Getting Ready, Step 3, as a model to simulate the motion of particles in the different states of matter. Tilt the pan slightly so that all the beads are in one area. Move the dish gently to model the solid state where particles vibrate but do not change positions relative to adjacent particles. Move it gently side to side until the beads are changing position but still in contact with each other. (This illustrates the liquid phase, in which particles still touch but adjacent particles can move relative to each other.) Finally, move the dish vigorously so the beads move throughout the dish. (This illustrates the gas phase, in which particles occupy the entire volume and are independent of the other particles.) The degree of shaking is analogous to the temperature; little shaking represents a lower temperature (and lower average kinetic energy of the particles), and more vigorous shaking represents a higher temperature (and higher average kinetic energy). Be sure that students observe the arrangement of the particles and the relative distances between them.

PROCEDURE

Part A: The Flick

1. Place the butane lighter on its side on the stage of an overhead projector. Gently shake the lighter from side to side so that all the students can observe the movement of liquid.

2. Ask, "Is the butane a solid, liquid, or gas?" *Liquid.*

3. Depress the valve and release some gaseous butane (but don't ignite the butane). Ask students what is happening.

4. Now, lower the lighter into the container of room-temperature water.

5. With the lighter on its side, depress the valve and ask the students to observe.

6. Ask, "Are you observing a solid, liquid, or gas? How can you tell?" *Gas bubbles escape.*

7. Determine the amount of gas escaping from the lighter using the method described in either Step 7a or 7b.

    a. Count the number of bubbles escaping from the lighter with its valve depressed in a 15-second period. Keeping the lighter on its side slows the rate at which the bubbles come out, thus making it easier to count them.

    b. Collect gas bubbles: Fill a test tube with water. Hold your finger over the end and invert. Submerge the tube and remove your finger. Position the opening of the tube above the lighter with its valve depressed, and collect gas bubbles for 15 seconds. (See Figure 2.) To ensure that the pressure of the gas inside the test tube is equal to atmospheric pressure, after the gas is collected, position the test tube so the level of water inside the tube is the same as the level of water in the container. (See Figure 3. If this is not done, comparison of the gas volumes from different trials will not be valid, since the gas pressures may differ in the trials.) Seal the container using a finger, remove it from the water, and measure the volume of water remaining in the container after the gas is collected. When this volume is subtracted from the volume of water in the full container, the volume of gas released is determined. (If a graduated cylinder is used, the volume may be read directly.)

 Depending on the lighter and the size of the test tube or graduated cylinder, you may need to increase or decrease the length of time for collecting bubbles.

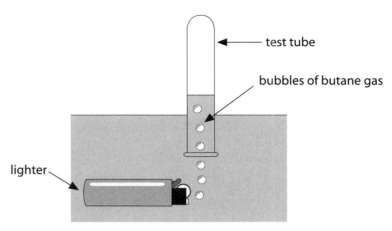

Figure 2: Collect gas bubbles in an inverted test tube.

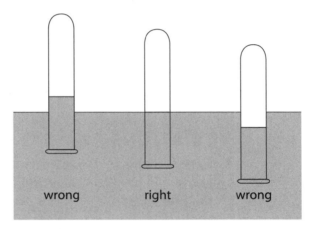

Figure 3: Ensure that the pressure in the test tube equals atmospheric pressure by lining up the water level inside the tube with the water level outside the tube.

8. Ask students to hypothesize how just a "flick" can change the state of butane.

9. After students have had a chance to speculate, tell them the freezing point (fp) and boiling point (bp) of butane. (At 1 atmosphere pressure, fp = −138.3°C and bp = −0.5°C.)

10. Ask a student to read a thermometer to determine the room temperature. Ask, "How does room temperature compare with the freezing point and boiling point of butane?" *Room temperature is higher than both.*

11. Challenge students to explain how butane can be found as a liquid at room temperature inside the lighter.

Part B: The Effect of Temperature on Bubble Rate

You may wish to have students record the gathered data on the Data Sheet (provided) as you perform the demonstration.

1. Measure the temperature of the room-temperature-water bath, the cold-water bath, and the warm-water bath and record the temperatures on the Data Sheet (provided).

2. Lower the butane lighter into the warm- or cold-water bath.

3. Depress the valve on the side of the lighter to release the gas as before (Part A, Step 7a or 7b) and measure the gas escaping in the same time period as used before.

4. Repeat Steps 2–3 with the other water baths.

5. Challenge students to explain any differences they may observe in the results at different temperatures.

6. Have students record the data from different trials, calculate averages (if duplicate trials were done), and make a bar graph of the data.

VARIATIONS AND EXTENSIONS

1. Have students investigate the following question at home: What household products in aerosol cans use butane or other hydrocarbons as the propellant?

2. Challenge students to speculate how they might change a gas to a liquid.

3. Show students an aerosol product that uses butane (or another hydrocarbon) as a propellant. Challenge students to hypothesize how the contents travel through the tube positioned inside the can.

EXPLANATION

The following explanation is intended for the teacher's information. Modify the explanation for students as required.

At room temperature (about 25°C or 77°F) and atmospheric pressure (1 atm or 760 torr), butane is a gas. When a pressure greater than 2.5 atm (as in the butane lighter) is applied to gaseous butane at 25°C, the butane condenses into the liquid state. A butane lighter contains liquid and gaseous butane. Pressing down on the lighter valve allows some butane to escape. Since the butane that comes out of the lighter is at atmospheric pressure, it is in the gaseous state. Butane in camp stove canisters is also under enough pressure that it exists mostly as a liquid at room temperature. If you move the canisters, you can hear the liquid sloshing inside. Likewise, the canisters for propane torches or gas barbecue grills contain propane mostly as a liquid.

Both pressure and temperature affect the state of matter. Many gases can be liquefied by increasing the pressure, even if the temperature of the gas is well above its normal boiling point. At the molecular level, increasing the pressure forces the gas molecules closer and closer together. When molecules get close enough, their tendency to behave independently is overcome by the strength of the intermolecular attractions, and the gas condenses into the liquid state.

In a butane lighter, liquid and gas butane both exist and are in equilibrium. That is, the rate at which liquid molecules are evaporating is the same as the rate at which gas molecules are condensing. The vapor pressure of a liquid is the pressure exerted by the gas molecules when gas and liquid are in equilibrium. The vapor pressure of a liquid depends upon the temperature; as the temperature increases, the vapor pressure increases. When the temperature of the butane lighter is increased, the vapor pressure of butane increases. This means that the pressure of butane gas (vapor) is increased, which in turn affects the rate at which butane is pushed out of the lighter.

Butane belongs to a group of organic compounds known as alkanes. All alkanes have the general formula of C_nH_{2n+2}. At room temperature and atmospheric pressure, the low-molecular-weight members of the group (methane, CH_4; ethane, C_2H_6; propane, C_3H_8; and butane, C_4H_{10}) are gases. Members of the group with larger numbers of carbon atoms are liquids or solids. Pentane (C_5H_{12}) is the first liquid in the series of linear alkanes, and octadecane ($C_{18}H_{38}$) is the first solid.

ASSESSMENT

Ask students, "What happens when the valve is opened on a propane tank (such as one used for fuel for a gas grill)?" Have students apply what they learned in this activity by writing a paragraph that answers the question on the particle level. Students should include drawings. Also, ask students, "Why are gases such as butane and propane commonly stored as liquids?"

CROSS-CURRICULAR INTEGRATION

Home, safety, and career:
*   Have students investigate the safe handling and storage of flammable materials.

REFERENCES

Marzzacco, C.J.; Speckhard, D. "Simple Demonstrations of the Liquefaction of Gases," *Journal of Chemical Education.* 1986, *63*(5), 436.

Shakhashiri, B.Z. *Chemical Demonstrations;* University of Wisconsin: Madison, WI, 1985; Vol. 2, pp 48–50.

CONTRIBUTORS

Alison Dowd, Talawanda Middle School, Oxford, OH; Teaching Science with TOYS peer writer.

Ann Hoffelder, Cumberland College, Williamsburg, KY; Teaching Science with TOYS, 1995.

Victoria Swenson, Grand Valley State University, Allendale, MI; Teaching Science with TOYS, 1995.

Janice Trumbull, Joy Elementary School, Fairbanks, AK; Teaching Science with TOYS, 1995.

HANDOUT MASTER

A master for the following handout is provided:
• Data Sheet

Copy as needed for classroom use.

Liquid to Gas in a Flick
Data Sheet

Directions: Record water temperature, collection time, and number of bubbles (or volume of gas) for each case.

	Warm Water	**Room-Temperature Water**	**Cool Water**
Water Temperature			
Collection Time			
Number of Bubbles Observed (Or Volume of Gas)			

Draw the setup used to collect the data in the space below.

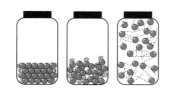

Disappearing Air Freshener

...Students discover how air freshener evaporates from a liquid gel to a gas.

The liquid perfume in the air freshener slowly evaporates from a liquid gel to a gas.

✔ Time Required

Setup	5	minutes
Performance	5–10	minutes once a week for 5 weeks
Cleanup	5	minutes

✔ Key Science Topics

- molecules and their properties
- physical changes
- states of matter

✔ Student Background

Students should be familiar with solids, liquids, and gases. It may be helpful if they have some prior experience working with graphs.

✔ National Science Education Standards

Science as Inquiry Standards:

- Abilities Necessary to Do Scientific Inquiry

  Students identify variables and make systematic observations.

  Students develop predictions, descriptions, and explanations using evidence from the air freshener investigation.

Physical Science Standards:

- Properties and Changes of Properties in Matter

  A characteristic property of the liquid gel used in air freshener is that its evaporation rate is affected by whether the system is open or closed, by surface area, and by temperature.

✔ Additional Process Skills

• predicting	Students predict what will happen to the air freshener samples.
• controlling variables	Students investigate variables that affect the rate of evaporation one at a time while holding others constant.
• collecting data	Students determine and record the masses of samples each week.

MATERIALS

For the Procedure

Per class or group
- 1 solid air freshener (such as Renuzit® or Glade®)
- balance
- 6 cupcake liners or pieces of filter paper
- sandwich-sized zipper-type plastic bag
- knife
- waxed paper or tin foil to put air freshener on while cutting
- goggles

For the Extension

Per group
- zipper-type plastic freezer bag, zipper-type plastic vegetable bag, or waxed paper bag

SAFETY AND DISPOSAL

No special safety or disposal procedures are required.

GETTING READY

1. Prepare two to three pieces of air freshener of approximately equal mass for each part of the investigation.

2. Place the air freshener for each investigation (1 piece for Investigation 1, 2 pieces for Investigation 2, and 3 pieces for Investigation 3) in a cupcake liner or on a piece of filter paper.

INTRODUCING THE ACTIVITY

Place the air freshener out of sight before students come into the room. Students will notice immediately that there is an aroma in the room. Ask, "What do you think is causing the odor? How do you know it's there? Can you find its location?" Show the air freshener to the class. Then discuss why it is called an air freshener. Have them observe its physical characteristics. Ask, "What state of matter does the air freshener in the container appear to be in? What state do you think it is in when you smell it in the room?" Give students a chance to share ideas, but do not provide answers at this time.

PROCEDURE

This activity includes three independent investigations. You may choose any or all of them. The class should be divided into groups accordingly.

Part A: The Investigation

1. Tell students that they will be investigating changes in the air freshener material under different conditions. Discuss what kinds of changes they might expect to observe in the air freshener over time and what variables might affect those changes.

2. Tell students that in this investigation they will be recording changes in mass of the air freshener under three different sets of conditions. Tell students to look over the "Observing the Air Freshener" Data Sheet (provided). Ask, "What is the variable in each of these investigations?" Have each group write a prediction of what will happen in each investigation on the Data Sheet. Instruct students to provide reasons for their predictions based on what they know about states of matter and changes of state.

3. Run each investigation for up to 5 weeks or as long as measurable changes are occurring. (See the Data Sheet for instructions.)

4. Have the students graph their group's data and share the results with other members of the class.

Part B: Class Discussion

1. What happened to the mass of each sample of air freshener over time? *It decreased.*

2. What caused the change in mass? *Water in the air freshener gel evaporated.*

3. For each of the investigations, review the variable. Have students share their initial predictions. Was the outcome different than they expected?

   Investigation 1   open system vs. closed system
   Investigation 2   surface area
   Investigation 3   temperature

4. Discuss the design of the air freshener container. Why is it adjustable? How does adjusting the position of the cover affect the amount of air freshener scent in the air?

5. If the Extension was done, discuss the effect that the type of closed system has on the loss of mass.

EXTENSION

To extend Investigation 1, have students explore the effect that using different types of bags has on the loss of mass. For example, have students try placing pieces of air freshener in zipper-type freezer bags, zipper-type bags for storing vegetables, or waxed paper bags. Have them compare the results with those from Investigation 1.

EXPLANATION

The following explanation is intended for the teacher's information. Modify the explanation for students as required.

Solid air fresheners generally consist of a gel of carrageenan and water, with a small amount of liquid fragrance (less than 5%) dispersed throughout the gel. Carrageenan is extracted from seaweed and is also used as a thickening agent for ice cream and other food products. The solid air fresheners work by a process of slow evaporation, allowing the dispersed liquid perfume to be released into the atmosphere at a controlled, slow rate. The gel also serves as an indicator for the consumer: when it has shrunk far enough, the solid air freshener is spent and it is time to buy a new one.

Investigation 1 involves comparing the rates of evaporation in an open system and a closed system. In an open system, gas particles can escape. In a closed system, both evaporation and condensation can occur. Initially, the rate of evaporation is greater, but as the number of vapor particles increases, the rate of condensation increases. Once the two rates are equal, no further observable change in the amount of gel will occur.

Investigation 2 involves comparing the rates of evaporation for different amounts of surface area. Because only surface molecules can evaporate, increasing the surface area by cutting the air freshener into smaller pieces increases the rate of evaporation.

Investigation 3 involves comparing rates of evaporation at various temperatures. The higher the temperature, the higher the rate of evaporation because more energy is available.

At first glance, it may appear that solid air fresheners sublime to the gas phase since no liquid phase is observed. As this Explanation describes, this is not the case. Examples of some solids that sublime at room temperature into the gas phase include moth balls and carbon dioxide (dry ice).

REFERENCES

Williams, B.; Doerhoff, B. *Challenges to Science: Physical Science;* McGraw-Hill: New York, 1979; pp 127–128.

Zaunbrecher, J. Air Care Products, World Wide Consumer Products Division, S.C. Johnson & Son, Inc., personal communication, 1996.

CONTRIBUTORS

Lynn Carlson, University of Wisconsin—Parkside, Kenosha, WI; Teaching Science with TOYS reviewer, 1996.

Marcia Denny, Anderson Community Schools, Anderson, IN; Teaching Science with TOYS, 1992–93.

HANDOUT MASTER

A master for the following handout is provided:
• Observing the Air Freshener—Data Sheet
Copy as needed for classroom use.

Disappearing Air Freshener
Observing the Air Freshener—Data Sheet

Investigation 1

1. Predict what will happen to the mass of a piece of air freshener sealed in a plastic bag in comparison to the mass of a sample exposed to air.

   _____

   _____

2. Record the mass of each air freshener sample on the table below under Week 1.

3. Place one sample in a zipper-type plastic bag and seal. Place the other sample on a cupcake liner and leave it exposed to the air.

4. Place both samples next to each other where they will not be disturbed.

5. Record the mass of each sample for 5 weeks using the following table.

		Mass of sample in grams			
Week Condition	1	2	3	4	5
Sealed in a bag					
Exposed to air					

6. Graph your results on a separate sheet of paper.

7. Explain your results on a separate sheet of paper, using what you observed as evidence.

Observing the Air Freshener—Data Sheet, page 2

Investigation 2

1. Predict what will happen to the mass of a piece of air freshener left in one piece ("uncut") in comparison to the total mass of a similar-sized sample that has been cut into small pieces ("cut-up").

_____

_____

2. Divide one of the samples into smaller pieces. Leave the other sample in one piece.

3. Record the mass of each air freshener sample (the uncut sample and the cut-up sample pieces) on the table below under Week 1.

4. Place the uncut sample on one cupcake liner and the cut-up sample pieces on another cupcake liner next to each other where they will not be disturbed. Be sure to place the cut-up sample pieces so they are spread out and not touching each other.

5. Record the mass of each sample for 5 weeks using the following table.

	Mass of sample in grams				
Week Condition	1	2	3	4	5
Whole					
Cut into pieces					

6. Graph your results on a separate sheet of paper.

7. Explain your results on a separate sheet of paper, using what you observed as evidence.

Observing the Air Freshener—Data Sheet, page 3

Investigation 3

1. Predict what will happen to the mass of a piece of air freshener that has been left at room temperature in comparison to a sample that has been near a heater and another sample that has been in a refrigerator.

_____

_____

2. Record the mass of each air freshener sample on the table under Week 1.

3. Place each sample on a cupcake liner. Place one sample near a heater and one sample in a refrigerator. Leave the third sample at room temperature.

4. Record the mass of each sample for 5 weeks using the following table.

Week Condition	1	2	3	4	5
Room Temperature					
Near heater					
Refrigerated					

Mass of sample in grams

5. Graph your results on a separate sheet of paper.

6. Explain your results on a separate sheet of paper, using what you observed as evidence.

Reproduced from *Investigating Solids, Liquids, and Gases with* **TOYS**, published by Terrific Science Press.

✔ *Hands-On Activity*

Demonstration

Learning Center

A Cool Phase Change

...This activity demonstrates that some liquids evaporate more readily than others and that a decrease in temperature accompanies evaporation.

✔ Time Required

Setup	10	minutes
Performance	25–30	minutes
Cleanup	5–10	minutes

✔ Key Science Topics

- evaporation
- volatility

✔ Student Background

This activity is most effective after students have discussed the three states of matter and phase changes.

Some liquids evaporate more rapidly than others.

✔ National Science Education Standards

Science as Inquiry Standards:

- Abilities Necessary to Do Scientific Inquiry
  Students make accurate measurements during a scientific investigation.
  Students use appropriate tools and techniques to gather, analyze, and interpret data.

Physical Science Standards:

- Properties and Changes of Properties in Matter
  A characteristic property of a liquid is its volatility, or speed of evaporation.

- Transfer of Energy
  Heat moves out of a liquid as the liquid evaporates.

✔ Additional Process Skills

• inferring	Students develop an idea of why the temperature decreases.
• making graphs	Students make bar graphs based on the data they collect.

MATERIALS

For Getting Ready
Per group
- masking tape and marker for labels
- dropper bottle for each test liquid
- the following test liquids:
  - nail polish remover containing either acetone or methyl ethyl ketone
  - 70% ethyl alcohol solution
  - 91–100% isopropyl alcohol
  - rubbing alcohol (70% isopropyl alcohol solution)
  - perfume
  - water
  - cologne
  - after-shave lotion
  - pre-shave lotion
  - oil
  - artificial or natural vanilla or other flavor extract
  - mouthwash

For Introducing the Activity
Per class
- cotton swabs
- rubbing alcohol

For the Procedure
Per class
- dropper bottles prepared in Getting Ready

Each group should receive a different liquid. Samples will subsequently be rotated between groups of students.

Per group
- alcohol or metal cooking thermometer (which reads between 0–30°C)

If additional thermometers are available, you may wish to give each group two thermometers and have them alternate between the two. This will cut down on the time students spend waiting for the thermometers to return to room temperature. (See Step 6 of the Procedure.)

- lump of clay
- cotton ball for each test liquid

SAFETY AND DISPOSAL

The liquids used in the Procedure are flammable, and their vapors are irritating to the eyes and respiratory system. Use these liquids only in a well-ventilated area and keep flames away. The liquids are harmful if ingested, and they can cause severe damage to the eyes. Should contact with the eyes occur, rinse with water for 15 minutes and seek immediate medical attention.

Certain liquids are inappropriate for use in the activity. For example, methanol and duplicating fluid are toxic when absorbed through the skin or when the fumes are inhaled. For this reason, use only the liquids suggested in Materials. Liquids can be saved for future use or flushed down the drain with water.

GETTING READY

Label and fill the dropper bottles with the liquids to be used.

INTRODUCING THE ACTIVITY

Rub a cotton swab dipped in rubbing alcohol and another dipped in water on the blackboard. Ask the students to observe what happens. Ask the students to guess which mark was made with alcohol, and ask them to give reasons for their guess. *Alcohol evaporates more quickly than water.*

Have students swab a little water and rubbing alcohol on their skin. Have them describe what they feel. Ask them to speculate why their skin feels cold. *Evaporation.* Which substance makes the skin feel colder? *Alcohol.* Point out that alcohol evaporates more quickly than water and also causes a more noticeable temperature change on your skin as it evaporates. Tell students that the faster a substance evaporates, the greater the temperature change that occurs. Ask, "How can we use this relationship to rank the evaporation speeds of various liquids?" Through class discussion, bring out the idea that students could measure the temperature change as different liquids evaporate.

PROCEDURE

Have each group do the following:

1. Wrap a cotton ball around the bulb end of the thermometer.

2. Prop the thermometer on a lump of clay so that the cotton does not touch the desk top.

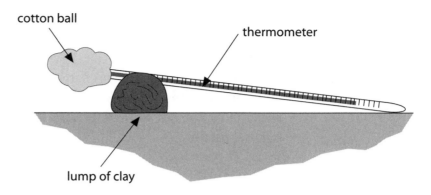

cotton ball

thermometer

lump of clay

Figure 1: Prop up the thermometer so the cotton ball is elevated.

3. Read and record the temperature of the thermometer.

4. Place 10–20 drops of the liquid onto the cotton ball. Record the number of drops actually used.

5. Carefully watch the thermometer to see if a temperature change occurs. Record the *lowest* temperature reading observed.

6. Remove the cotton ball and allow the thermometer to return to room temperature.

7. Using a new cotton ball for each liquid, repeat Steps 1–6 using the same number of drops of each of the remaining liquids as used for the first liquid.

8. Calculate the change in temperature for each liquid tested and record the calculations on the "Temperature Changes of Liquids" Data Sheet (provided). Prepare a bar graph representing the change in temperature for each sample. (See Figure 2 for a sample bar graph.)

Figure 2: Temperature Changes in Liquids

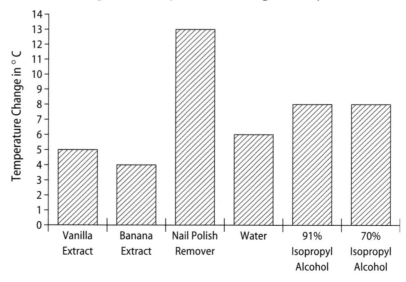

9. Ask, "Which liquid evaporated the fastest? Why do you think so?" *Nail polish remover evaporated fastest. We could tell because it caused the greatest drop in temperature.*

EXPLANATION

The following explanation is intended for the teacher's information. Modify the explanation for students as required.

Heat is a form of energy that causes the motion of the particles that make up matter. The temperature of a sample of liquid is related to the average energy of the particles in the sample. Forces of attraction between these particles hold the particles in the liquid together. If a particle moving near the surface of a liquid has enough energy, it may overcome the attractive forces of the liquid and escape. This process is called evaporation. The rate of evaporation in liquids depends on several factors, including the mass of particles, temperature, and the strength of the intermolecular forces (attractions between particles). Liquids that evaporate quickly are said to be volatile. Since particles escaping to the gas phase are surface particles with higher energies, the average energy of the liquid particles remaining decreases, and the temperature of the remaining liquid decreases. The more volatile the liquid, the greater the observed temperature change. Table 1 provides sample data for some of the liquids your students may have tested in this activity. (The bar graph in Figure 2 is based on this data.)

Table 1: Typical Temperature Changes for Some Liquids			
Liquid	Room Temperature	Lowest Temperature	Temperature Change
vanilla extract	25°C	20°C	–5°C
banana extract	25°C	21°C	–4°C
nail polish remover	25°C	12°C	–13°C
water	25°C	19°C	–6°C
91% isopropyl alcohol	25°C	17°C	–8°C
70% isopropyl alcohol	25°C	17°C	–8°C

CROSS-CURRICULAR INTEGRATION
Life science:
* Discuss how the process of sweating during exercise helps the body.
* Discuss why alcohol rubs are used to reduce fever.
* Discuss why dogs and other animals pant when they are hot.

REFERENCE

Murphy, J.; Zitzewitz, P.; Smoot, R. *Physics: Principles and Problems;* Merrill: Columbus, OH, 1986.

CONTRIBUTORS

Lorinda Bagshaw, Lawrenceburg High School, Lawrenceburg, IN; Teaching Science with TOYS, 1992–93.

Diana Klenk, Institute for Chemical Education participant, Madison, WI.

HANDOUT MASTER

A master for the following handout is provided:
*   Temperature Changes of Liquids—Data Sheet

Copy as needed for classroom use.

Name _____   Date _____

A Cool Phase Change
Temperature Changes of Liquids—Data Sheet

1. Record the experimental data and calculate the change in temperature for each liquid tested.

Number of drops of liquid used: _____

Temperature Changes of Liquids			
Liquid	Room Temperature	Lowest Temperature	Temperature Change

2. In the space below, prepare a bar graph representing the change in temperature for each liquid.

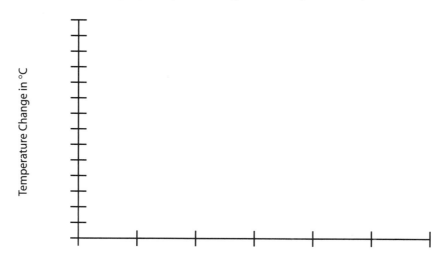

Liquids

Using Dry Ice to Inflate a Balloon

...Students observe the large volume change associated with sublimation.

✔ Time Required

Setup   5 minutes
Performance   5–15 minutes*
Cleanup   5 minutes
*Additional time is required if done as a hands-on activity.

✔ Key Science Topics

- sublimation
- volume change

✔ Student Background

This activity is most effective if students already know that dry ice is solid carbon dioxide and that carbon dioxide is a gas at room temperature and pressure.

As the dry ice inside the sealed balloon sublimes, the balloon inflates.

✔ National Science Education Standards

Science as Inquiry Standards:

- Abilities Necessary to Do Scientific Inquiry

  Students think critically about evidence and form a logical argument about the cause-and-effect relationship in the experiment.

Physical Science Standards:

- Properties and Changes of Properties in Matter

  A characteristic property of gases is that they occupy a much greater volume than an equivalent mass of the solid.

✔ Additional Process Skills

- observing   Students observe as the balloon containing carbon dioxide inflates.
- predicting   Students predict what will happen when dry ice is placed in water and in a sealed balloon.

213

MATERIALS

For Getting Ready
Per class
- dry ice [about 250 g (½ pound), 3 inches x 3 inches x 1 inch]
- gloves
- goggles
- insulated Styrofoam™ container or ice chest with a lid (for storing the dry ice)

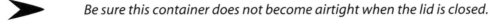 *Be sure this container does not become airtight when the lid is closed.*

- hammer (to crush the dry ice)

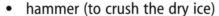 *If possible, purchase the dry ice the day it is to be used. Only a small amount of dry ice is needed, but the quantity will shrink as the dry ice sublimes, and you'll need enough to last until it is used. The amount listed is the amount needed to conduct the experiment. If the dry ice will not be used immediately, get enough so the amount listed will be available when you begin doing the activity. Check the Yellow Pages for a source of dry ice or inquire at local ice cream stores. Dry ice may be difficult to obtain, so identify a source before planning to do this activity.*

For Introducing the Activity
Per class
- small pieces of dry ice
- container of warm water
- gloves or tongs
- goggles

For the Procedure
Per student or group
- round balloon
- spoonful of crushed dry ice

Per class
- spoon
- pencil
- funnel

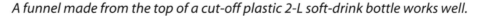 *A funnel made from the top of a cut-off plastic 2-L soft-drink bottle works well.*

Per student
- gloves, mittens, old socks, or a cloth towel
- goggles

For the Extension
Per class
- a large plastic garbage bag and tie
- 44 g dry ice

SAFETY AND DISPOSAL

Dry ice is cold enough to cause frostbite or blisters. Use gloves or tongs and wear safety goggles whenever handling dry ice. Use in a well-ventilated area. Be sure the activity is carried out as written. Allow unused dry ice to sublime in a well-ventilated area.

GETTING READY

1. Dry ice can be stored for short periods of time in a polystyrene foam (such as Styrofoam) chest. Do not break up big pieces until just before use in order to increase the length of time the dry ice will last.

2. Wearing goggles and gloves or other hand protection, use a hammer to crush the dry ice into pieces small enough to fit through the funnel; the smaller the pieces, the faster the balloon will inflate.

The amount of dry ice you need depends on the size of balloon you are using.

3. Too much dry ice will pop the balloon. Experiment with different amounts of dry ice in the balloon before doing the activity with students.

INTRODUCING THE ACTIVITY

Place a few small pieces of dry ice on the table. Ask the students if they know what this is and why it is called "dry." *Dry ice; it is cold but does not melt to a liquid like water ice does.* Ask, "If the dry ice doesn't melt, what happens to it when it sits out for a while?" *It sublimes into gaseous carbon dioxide (CO_2), which disperses into the air.* Point out that dry ice is very cold (about $-78°C$) and that hand protection is needed to handle it or it can cause frostbite. Also point out that CO_2 is a colorless, odorless gas; thus the white cloud that appears to be coming from the dry ice is not CO_2 but actually small droplets of water that are formed as water vapor in the air is condensed by the cold, gaseous CO_2. Likewise, on humid days, some frost or moisture from condensed water vapor may remain after all the dry ice has sublimed.

Ask the students to predict what would happen if you place some of the dry ice into a container of warm water. *The dry ice will sublime.* Do this and point out that the bubbles are full of gaseous CO_2. Remind them that the clouds are condensed water vapor.

Ask students to predict what would happen if you placed some dry ice into a balloon and tied it off.

PROCEDURE

1. Stretch the neck of the balloon over the end of the funnel as shown in Figure 1.

Figure 1: Stretch the balloon over the end of a funnel.

2. Using gloves, mittens, old socks, or a cloth towel, place about half of a spoonful of dry ice into the funnel and use a pencil to push the dry ice into the balloon.

The amount of dry ice you need depends on the size of balloon you are using—too much will pop the balloon and can result in dry ice spraying around the room. If the balloon seems to be expanding so much that it may pop, you can pop the balloon yourself by cutting it very near the tie—this can be done inside a waste basket or other protected area.

3. Tie off the balloon and shake it until all the dry ice sublimes. Ask students to observe carefully. Ask, "What evidence do you have that all the dry ice has sublimed?" *The balloon has inflated and the sound of solid material knocking around is gone.*

EXTENSION

Determine the approximate volume of 1 mole carbon dioxide (CO_2) gas as follows: Place 44 grams dry ice (1 mole CO_2) in a large garbage bag, squeeze as much of the air as possible out of the bag, seal the bag with a twist-tie, and allow the dry ice to sublime. Assuming no CO_2 gas leaks out of the bag, the final volume will be that of 1 mole CO_2 gas, about 24 L at room temperature; the exact volume depends on the temperature and pressure. Discuss the difference in volume between 1 mole solid CO_2 and 1 mole CO_2 gas.

EXPLANATION

The following explanation is intended for the teacher's information. Modify the explanation for students as required.

Dry ice is solid carbon dioxide (CO_2). It sublimes (that is, changes directly from the solid state to the gaseous state) at −78°C. When dry ice is placed in the balloon and the balloon is tied off, the dry ice changes to gaseous CO_2, which occupies a much greater volume than the equivalent mass of solid, and the balloon expands. The balloon continues to expand until all of the dry ice has sublimed or until the balloon ruptures, whichever occurs first.

The Extension provides a way of visualizing the volume of 1 mole CO_2 gas. Some of the CO_2 may leak from the plastic bag, and the conditions will not exactly be standard temperature and pressure (0°C and 1 atm). The result, however, does provide a demonstration of the much larger volume of a mole of a gas as compared with a mole of a solid or liquid.

ASSESSMENT

Have the students draw a graphic representation of what happened to the molecules of carbon dioxide (CO_2) in this phase change.

CROSS-CURRICULAR INTEGRATION

Home, safety, and career:
- Have students research the process of freeze-drying, a process that involves the sublimation of water.

Social studies:
- Research how clothes were dried in the cold regions of the world before automatic dryers.

REFERENCE

"A Balloon Full of Carbon Dioxide"; *Fun with Chemistry: A Guidebook of K–12 Activities;* Sarquis, M., Sarquis, J., Eds.; Institute for Chemical Education: Madison, WI, 1993; Vol. 1, pp 13–15.

The Phase Changes of Carbon Dioxide

...Students discover that solid carbon dioxide (dry ice) sublimes under normal conditions but will melt at high pressures.

✔ Time Required

Setup	15 minutes
Performance	10 minutes*
Cleanup	5 minutes

*Additional time is required if done as a hands-on activity.

Melting dry ice in a pipet

✔ Key Science Topics

- sublimation
- phase changes
- states of matter

✔ Student Background

The students should be familiar with the three states of matter and phase changes.

✔ National Science Education Standards

Science as Inquiry Standards:

- Abilities Necessary to Do Scientific Inquiry

  Students make systematic observations as they observe changes of state of carbon dioxide.

  Students differentiate description from explanation and provide explanations based on evidence and logical argument.

  Students listen to and respect the explanations proposed by other students.

  Students communicate observations and tell other students about their explanations.

Physical Science Standards:

- Properties and Changes of Properties in Matter

  A characteristic property of carbon dioxide is that it changes from solid to gas at normal atmospheric pressure but changes to a liquid only at pressures above 5.1 atmospheres.

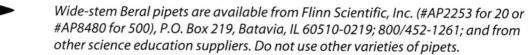

MATERIALS

For Getting Ready
Per class
- scissors
- 10-ounce clear plastic cup
- quart-sized zipper-type plastic bag
- rubber band
- cloth towel and hammer or a mortar and pestle
- 3–4 wide-stem Beral pipets

➤ *Wide-stem Beral pipets are available from Flinn Scientific, Inc. (#AP2253 for 20 or #AP8480 for 500), P.O. Box 219, Batavia, IL 60510-0219; 800/452-1261; and from other science education suppliers. Do not use other varieties of pipets.*

- insulated Styrofoam™ container or ice chest with a lid (for storing the dry ice)

➤ *Be sure this container does not become airtight when the lid is closed.*

- newspapers (for additional insulation)
- gloves or tongs
- goggles
- about 250 g (½ pound) dry ice, 3-inch x 3-inch x 1-inch block

➤ *If possible, purchase the dry ice the day it is to be used. Only a small amount of dry ice is needed, but the quantity will shrink as the dry ice sublimes, and you'll need enough to last until it is used. The amount listed is the amount needed to conduct the experiment. If the dry ice will not be used immediately, get enough so the amount listed will be available when you begin doing the activity. Check the Yellow Pages for a source of dry ice or inquire at local ice cream stores. Dry ice may be difficult to obtain, so identify a source before planning to do this activity.*

For Introducing the Activity
Per class (or per group if done as a hands-on activity)
- water ice cube
- piece of dry ice about the same size as the water ice cube
- clear container of water
- gloves or tongs
- cup
- candle
- matches

For the Procedure
Per student
- goggles

Per class (or per group if done as a hands-on activity)
- pipets prepared in Getting Ready
- bag-cup setup prepared in Getting Ready
- powdered dry ice prepared in Getting Ready
- pliers

SAFETY AND DISPOSAL

Dry ice is cold enough to cause frostbite or blisters. Use gloves or tongs and wear safety goggles whenever handling dry ice. Use in a well-ventilated area. Be sure the activity is carried out as written. The sealed Beral pipet containing dry ice should be submerged in water to help absorb the impact of the explosion. DO NOT use pipets other than the wide-stem pipets listed in Materials. Allow unused dry ice to sublime in a well-ventilated area.

GETTING READY

1. Use scissors to cut off one-third of the stem of a wide-stem Beral pipet. (Remove the tapered portion as shown in Figure 1.) The dry ice will be placed in this pipet.

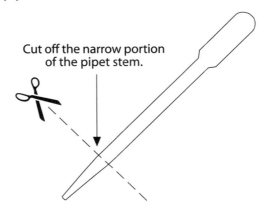

Cut off the narrow portion of the pipet stem.

Figure 1: Prepare the Beral pipet.

2. Fill the plastic cup half-full with tap water. Place the plastic bag over the rim of the cup and place a rubber band just below the rim to hold the bag on the cup. Cut a 1-inch vertical slit into one side of the bag just above the rim of the cup. The pipet will be inserted through the slit and into the water in the cup. (See Figure 2.)

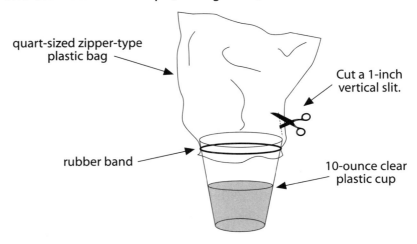

quart-sized zipper-type plastic bag

Cut a 1-inch vertical slit.

rubber band

10-ounce clear plastic cup

Figure 2: Place a plastic bag over a cup half-full of water.

3. Powder the dry ice by wrapping it in the cloth towel and gently hitting it with the hammer or by using a mortar and pestle.

 To keep the rate of sublimation slow, don't crush the dry ice until just before use. Store the dry ice in an insulated container (or keep it wrapped in newspaper).

INTRODUCING THE ACTIVITY

Place a regular (water) ice cube and a piece of dry ice of about the same size in a location that can be observed by all students. Have students make observations. The ice cube melts, but the dry ice just seems to disappear! Ask students where the dry ice is going. They may say that the white vapor drifting away from the dry ice is gaseous carbon dioxide (CO_2), but tell them that gaseous CO_2 is colorless. Ask, "What do you think the white vapor is?" If necessary, guide the discussion by reminding students that they sometimes see a similar white vapor when they exhale. *When you exhale outside on a very cold day, a white fog appears.* Why doesn't this fog appear on warm days? Lead students to conclude that the fog when they exhale and the fog around the dry ice is condensed water vapor.

When you have established that the white mist is not CO_2, discuss what is actually happening to the dry ice. Tell students that the dry ice sublimes (changes from the solid state to the gas state directly) without melting into the liquid state. This phase change is in contrast to water ice, which melts to liquid water. The cloud-like substance drifting away from the surface of the dry ice is condensed water vapor (very small droplets of water that are suspended in air). The water vapor in the air condenses when it comes into contact with the cold dry ice and with the water ice cube. (If it is a humid day, water vapor in the air will condense and freeze on the surface of the dry ice. When these fluffy crystals of ice are knocked off, they will melt and leave a small puddle, but the amount of liquid will be much less than that from the melting ice cube.)

Ask students, "If CO_2 vapor is colorless, how can we be sure that the dry ice is actually changing to the gas state? In other words, what could we do to test for the presence of this gas?" Test any feasible suggestions. Two possibilities are as follows: Drop a piece of dry ice into a clear container of water. Have students continue making observations. Students should observe that bubbling occurs as the solid disappears. If some dry ice is allowed to sublime in a cup, the gaseous CO_2 can be poured over a candle flame. When the candle goes out, identify the gas as CO_2.

Ask students which two states of matter they have observed so far for the CO_2. *Solid and gas.* Which state have they not observed for the CO_2? *Liquid.* Discuss what conditions might affect the state of CO_2 or any other substance. Students will probably say "temperature," based on common experiences with substances such as water, but lead them to recognize that pressure also affects the state of matter. If the pressure is below 5.1 atmospheres (atm), CO_2 will never be in the liquid phase, no matter how low the temperature is. Tell them that they will be able to observe CO_2 in liquid form in this activity. Ask them to consider all that they have learned about the states of matter to figure out how the high pressure required for liquefaction is produced.

PROCEDURE

1. Discuss the importance of making systematic observations in a scientific investigation and the difference between description and explanation. Emphasize that during the demonstration students should be observing carefully and making clear descriptions on their Observation Sheets (provided). Warn students that changes may happen quickly and that you may need to go to the next step before they are finished recording observations. They can fill in any missing observations after the demonstration is complete. You will then ask them to use their observations to explain what happened.

2. Place powdered dry ice in the wide-stem pipet by scooping it into the tip. Invert the pipet so the dry ice falls into the bulb. (See Figure 3.) Repeat this until the pipet bulb is about one-quarter full. Give students a moment to observe and record descriptions on their Observation Sheets.

 If dry ice sticks to the stem of the pipet, gently tap or flick the pipet until all the dry ice is in the bulb.

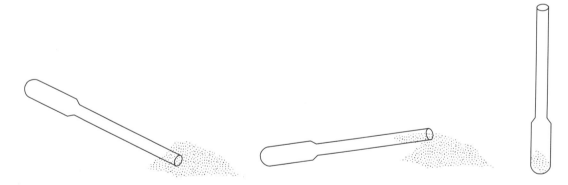

Figure 3: Scoop powdered dry ice into the pipet.

 When the system is clamped shut with the pliers in the next step, the pressure will increase as the CO_2 sublimes. Once the CO_2 is liquefied, the pipet will often burst with a moderate to loud "pop." Read Steps 3–8 before proceeding so you can anticipate what will happen.

3. Fold over 1 cm of the stem of the pipet. Using a pair of pliers, clamp the stem securely shut so that no gas can escape. (See Figure 4.)

The longer the handles on the pliers, the more easily you'll be able to seal the pipet tightly so no gas escapes.

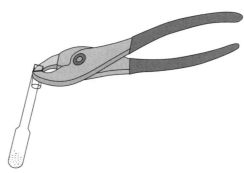

Figure 4: Clamp the pipet stem securely shut.

4. Keeping the pipet stem tightly clamped, slide the pipet through the slit in the plastic bag and submerge the pipet bulb in the cup of water. (See Figure 5.) Have students observe the behavior of the dry ice and record descriptions on the Observation Sheet (through the changes described in Steps 5–8).

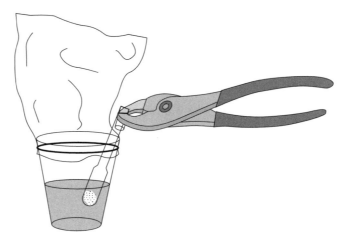

Figure 5: Slide the clamped pipet through the slit in the plastic bag.

5. After a while, the dry ice will liquefy. (If it doesn't liquefy, the pipet bulb is not sealed.) Remind students that CO_2 will liquefy only at pressures above 5.1 atm. Ask students, "How did the pressure in the pipet increase?" The grip with the pliers should not have changed, and the pipet should not have been squeezed any differently than it was at first. *As the CO_2 sublimes, the gas exerts pressure inside the pipet. When the pressure exceeds 5.1 atm, the remaining solid CO_2 begins to melt (become a liquid).* As soon as the dry ice melts, carefully loosen the grip

on the pliers (still holding the bulb in the water) so some gaseous CO_2 escapes, and observe the CO_2 solidify. Ask students, "Why is the CO_2 turning back to the solid state?" *The pressure is decreasing below 5.1 atm.* Ask students what caused the decrease in pressure. *When the pliers were loosened, some gas escaped.*

This takes a lot of practice; if the pipet remains sealed too long, the pressure increases enough to rupture the pipet, as described in Step 8. With practice, you can release some gas and decrease the pressure before the pipet ruptures. As you lead discussion, keep an eye on the pipet; you may not be able to complete each stage of discussion before the next phenomenon occurs. In this case, have students observe carefully and discuss after this step is completed.

6. Tighten the grip on the pliers again and observe.

7. Repeat Steps 5 and 6 several times.

8. Still holding the bulb in the water, tighten the grip on the pliers again. Observe. (The pipet will burst.) Ask students, "What caused the pipet to pop?" *The pressure exerted by the gaseous CO_2 increased so much that the pipet ruptured.*

The pipet bursts with a pop. The intensity of the pop varies from one trial to the next, depending on the integrity of the bulb. It rarely exceeds the noise level of the pop of a typical balloon. The pipets, made from low-density polyethylene, do not fragment, shatter, or produce sharp edges upon rupturing. When the bulb does explode, it actually tears open suddenly at its weakest spot. The water helps to muffle the sound and also serves to catch any small pieces of dry ice that might be expelled. The bag over the cup decreases the amount of splashing when the pipet bulb ruptures.

9. If desired, repeat Steps 2–8 so that students can review their Observation Sheets and fill in any missing descriptions.

10. As a class, discuss students' observations and explanations. Encourage them to listen to and respect the explanations proposed by other students. Through discussion, clarify any misconceptions.

11. Have students work individually or in small groups to complete the Question Sheet (provided) and discuss as a class.

EXPLANATION

The following explanation is intended for the teacher's information. Modify the explanation for students as required.

Carbon dioxide (CO_2) is an important industrial compound. It is also important to living systems; it is used by plants in photosynthesis and produced in respiration. CO_2 exists in the gaseous state at normal room conditions of approximately 1 atm pressure and 20°C. Under different pressure and temperature conditions, CO_2 can exist as a liquid or a solid. Many of us are familiar with the solid form of CO_2, dry ice. At 1 atm pressure and temperatures of -78°C (-108°F) or below, CO_2 exists as dry ice. At room temperature and atmospheric pressure, solid CO_2 sublimes spontaneously from solid to gas, thus earning its nickname of "dry ice," since it does not melt to a liquid the way ordinary ice does.

CO_2 is less commonly seen in the liquid state because it exists as a liquid only at pressures above 5.1 atm. If the pressure is below 5.1 atm, CO_2 will never be in the liquid phase, no matter what the temperature is. In this activity, when the dry ice sublimes in the sealed pipet, the pressure increases. When the pressure exceeds 5.1 atm, CO_2 can exist in the liquid state, and the solid begins to melt. When the pliers are loosened, if some of the gas escapes, the pressure is lowered, and the liquid CO_2 refreezes to form dry ice.

A phase diagram helps to explain the liquefaction of CO_2. (A phase diagram summarizes the states of matter that exist at specific temperatures and pressures.) Figure 6 shows the phase diagram for CO_2. For example, at a pressure of 1 atm and a temperature below -78°C, CO_2 exists as a solid; above -78°C, it exists as a gas. Line A in Figure 6 represents the sublimation of solid CO_2 at 1 atm as it is warmed above -78°C. This represents the usual change of state for solid CO_2 at normal pressure. Line B in Figure 6 shows that at a pressure above 5.1 atm, solid CO_2 will melt into a liquid when its temperature increases.

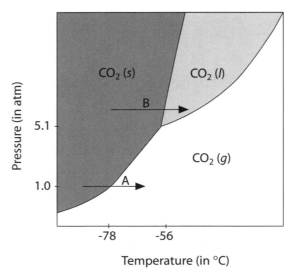

Figure 6: Phase change diagram for carbon dioxide

The properties of CO_2 make it useful for a variety of applications, one of the most familiar of which is the CO_2 fire extinguisher. These fire extinguishers contain liquid CO_2 as well as gaseous CO_2. When the fire extinguisher handle is squeezed, CO_2 escapes in the gas phase, since it is at atmospheric pressure once it comes out of the fire extinguisher. This corresponds to Line C in Figure 6.

Investigating Solids, Liquids, and Gases with **TOYS**

ASSESSMENT
Options:

- Have students complete the Observation Sheet (provided).

- Have students complete the Question Sheet (provided).

- Have students complete the Assessment Sheet (provided).

CROSS-CURRICULAR INTEGRATION
Home, safety, and career:
- Sublimation is used to "freeze-dry" foods. Have students find examples of foods that were prepared by freeze-drying.
- Liquid CO_2 is now being used as an agent for extracting oil from shale, as a medium for disposal of hazardous waste, as a dry cleaning solvent (replacing more-dangerous solvents), and to extract caffeine from coffee to make "decaffeinated" coffee. Dry ice is also used to seed clouds to promote rainfall on a small scale. Have students research these uses of CO_2.

REFERENCES

Becker, R. "Wet Dry Ice," *Journal of Chemical Education*. 1991, *68,* 782.

Brown, T.L.; LeMay, H.E.; Bursten, B.E. *Chemistry, The Central Science,* 6th ed.; Prentice Hall: Englewood Cliffs, NJ, 1994; p 392.

CONTRIBUTOR

Lee Ann Ellsworth, Franklin Middle School, Springfield, OH; Teaching Science with TOYS fellow, 1995.

HANDOUT MASTERS

Masters for the following handouts are provided:
- Observation Sheet
- Question Sheet and Answer Key
- Assessment Sheet

Copy as needed for classroom use.

Name _____   Date _____

The Phase Changes of Carbon Dioxide
Observation Sheet

For each step of the Procedure, draw the contents of the pipet, and describe what you see. After the Procedure is completed, write your explanation for what happened in each step.

Draw the contents of the pipet.	Describe what you see.	Explain what happened.

unsealed pipet

when first sealed

after pressure builds up

Reproduced from *Investigating Solids, Liquids, and Gases with* **TOYS**, published by Terrific Science Press.

Draw the contents of the pipet.	**Describe what you see.**	**Explain what happened.**

when pliers are loosened

| | _____ | _____ |

when kept sealed longer

The Phase Changes of Carbon Dioxide
Question Sheet

1. What do you observe when the dry ice sublimes? What do you observe when it melts?

    _____

    _____

    _____

2. How is the melting of dry ice different than the melting of ordinary ice? What ideas can you offer to explain these differences?

    _____

    _____

    _____

    _____

3. As you melt and refreeze the CO_2 sample over and over, why does it eventually get used up?

    _____

    _____

    _____

    _____

4. What might have happened if less dry ice had been placed inside the pipet bulb? Why?

    _____

    _____

    _____

5. What happened when the grip on the pliers was not released after the dry ice melted? Why?

    _____

    _____

1. When the dry ice sublimes, a misty, smoke-like substance is produced. This mist appears because water vapor from the air condenses into a fog (small droplets of water suspended in air) as the air is cooled by the cold CO_2. When the dry ice melts, it turns into a clear, colorless liquid.

2. An increased pressure is required for dry ice to melt. The melting of ordinary ice at atmospheric pressure and dry ice at elevated pressure appear to be similar. However, one visible difference is that the solid CO_2 remains at the bottom of the liquid, whereas ordinary ice floats in water. This suggests that solid CO_2 is more dense than liquid CO_2. Water is quite unusual in that its solid form is less dense than its liquid form.

3. Some of the CO_2 in the pipet must sublime into a gas in order to build up enough pressure to liquefy the rest. Each time the pressure is released to refreeze the sample, some gaseous CO_2 escapes from the pipet.

4. The sample may not have liquefied. If less dry ice is used, the sample could sublime completely without building up enough pressure to reach the 5.1 atm necessary to liquefy CO_2.

5. When the grip on the pliers is not released after the dry ice melts, the bulb ruptures because the pressure continues to increase.

Name _____   Date _____

The Phase Changes of Carbon Dioxide
Assessment Sheet

Identify the state or states of CO_2 that could be present in each case illustrated in the figures below. Explain your answers.

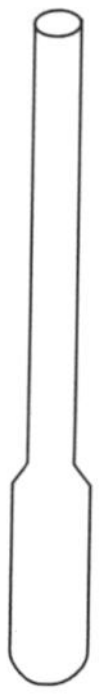

unsealed pipet

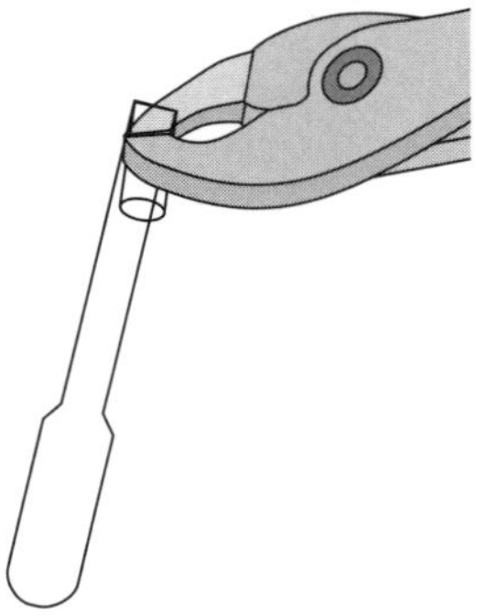

when first sealed

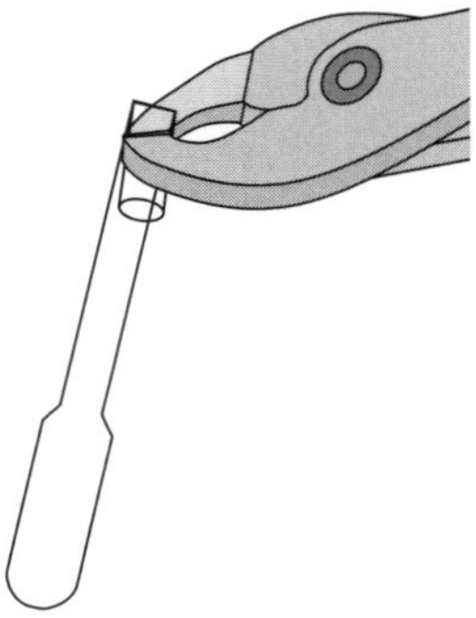

after pressure builds up

_____

_____

_____

state(s)

_____

_____

_____

state(s)

_____

_____

_____

state(s)

explanation

_____

_____

_____

_____

_____

_____

_____

explanation

_____

_____

_____

_____

_____

_____

_____

explanation

_____

_____

_____

_____

_____

_____

_____

Balloon-into-a-Flask Challenge

...Students observe the dramatic effect that the steam-to-water phase change has on volume.

✔ **Time Required**

Setup	10–15	minutes
Performance	20–30	minutes
Cleanup	5–10	minutes

✔ **Key Science Topics**

- effect of temperature on the volume of a gas
- phase change
- properties of gases

✔ **Student Background**

This activity works best if students already understand that air, an invisible gas, takes up space. They should also have some experience with the basic concepts of phase changes, particularly as related to boiling and condensation of water.

The balloon is pushed into the cooling flask.

✔ **National Science Education Standards**

Science as Inquiry Standards:

- Abilities Necessary to Do Scientific Inquiry

  Students differentiate description from explanation.

  Students review data from the demonstrations, summarize the data, and form a logical argument about the cause-and-effect relationships.

  Students communicate observations.

Physical Science Standards:

- Properties and Changes of Properties in Matter

  A characteristic property of gases is that they expand when heated.

  A characteristic property of steam is that it occupies a much larger space than an equal mass of liquid water.

MATERIALS

For Introducing the Activity
Per class

- medium-sized Erlenmeyer or Florence flask (125- to 500-mL)

Do not substitute another container, because of the danger of shattering when heated.

- balloon
- Balloon-into-a-Flask setup prepared in Getting Ready

For the Procedure
Per class

- medium-sized Erlenmeyer or Florence flask (125- to 500-mL)

Do not substitute another container, because of the danger of shattering when heated.

- 1–2 balloons
- 1 of the following heat sources:
    - hot plate
    - Bunsen burner with ring stand, iron ring, clamp, and wire gauze
- container for boiling water
- tongs
- second pair of tongs or heat-resistant gloves
- goggles
- (optional) ice water in a dishpan
- (optional) metal cooking or alcohol thermometer
- (optional) 10–20 drops of a soapy water solution
- (optional) dropper

For the Extension
Per class
- egg-sized wad of steel wool
- vinegar
- Erlenmeyer or Florence flask (125- to 500-mL)
- balloon

SAFETY AND DISPOSAL

Use caution with flames and hot flasks. Steam escaping from the flask can cause severe burns. Because of potential breakage of thermometers, metal cooking or alcohol thermometers should be used rather than mercury thermometers. No special disposal procedures are required.

GETTING READY

For Introducing the Activity, prepare a Balloon-into-a-Flask setup by following Steps 1–7 of Part D of the Procedure.

INTRODUCING THE ACTIVITY

Carry out the following exercise to provide a basis for student understanding of the feat accomplished in the activity:

1. Show the class a balloon and a flask. Ask them to suggest ways of getting the balloon into the flask. *A typical response will be to push the balloon into the flask so that it falls to the bottom.* Do this to show that this is one way to put the balloon inside the flask. Ask the students why the balloon was able to go into the flask. *The flask was open so the balloon could slip in; there was nothing to prevent the balloon from going into the flask.*

2. Now add a new restriction to the problem: stretch the mouth of the balloon over the flask opening as shown in Figure 1. Ask the students to suggest ways of getting the balloon into the flask without removing the balloon from the mouth of the flask. Some may again suggest trying to push the balloon into the flask. Try this to show them that it does not work. Ask students why the balloon could not be pushed into the flask as before. *The air in the flask takes up space and prevents the balloon from occupying the same space.*

3. Show the students the Balloon-into-a-Flask setup prepared in Getting Ready. Ask students to suggest an explanation for how the balloon got into the flask. Use their suggestions to lead into the Procedure.

PROCEDURE

Part A: Organize the Observations

1. Tell students that they will be observing three methods for attempting to get a balloon into a flask. Discuss the importance of making systematic observations during a scientific investigation so that the evidence collected can be used to propose explanations for the observations. Through discussion, clarify the difference between describing observations and proposing explanations.

2. Show students the Observation Sheet (provided) and explain that during the demonstrations, you want them to draw their observations.

Part B: Heat the Flask with the Balloon Attached

1. Prepare the balloon by stretching it several times. (This allows the balloon to be more easily stretched over the lip of the flask in Step 2.)

2. Stretch the mouth of the balloon over the flask opening. (See Figure 1.) Ask the students what is inside the balloon. *Air.*

Figure 1: Stretch a balloon over the opening of a flask.

 Be careful not to burn yourself with steam while heating the flask. Take care not to let the balloon expand to too large a size, which might cause it to pull off the neck of the flask or pop.

3. Tell the students to write down what they see as the flask is heated. Heat the flask by holding it with a clamp or tongs in a container of boiling water.

4. Remove the flask from the boiling water and allow it to cool to room temperature.

 The balloon will collapse to its original position but will not be pushed into the flask. To speed up the cooling process and therefore the collapsing action, place the flask in ice water. Do not leave the flask in the ice water too long or you may lower the temperature below room temperature and cause the balloon to be slightly pushed into the flask.

5. (optional) You can demonstrate that the expansion and contraction processes can be repeated by returning the flask to the boiling water and warming carefully as before, then cooling.

 This procedure can usually be repeated a number of times; however, the balloon will be weakened by the stretching and contracting process and may develop a leak or tear.

Part C: Heat the Flask and Then Put the Balloon On

1. Prepare the balloon by stretching it several times.

 Be careful not to burn yourself with steam while heating the flask.

2. Heat the flask by holding it with a clamp or tongs in the boiling water for about 5 minutes to be sure the air inside the flask has been thoroughly heated. Have students write down what they see during this heating process.

3. (optional) Measure and record the temperature of the heated air.

4. Remove the flask from the heat source. Have an assistant use a clamp, heat-resistant gloves, or tongs to hold the flask steady while you quickly but CAREFULLY stretch the mouth of the balloon over the flask opening. (Using a clamp is the preferred method. See Figure 2.)

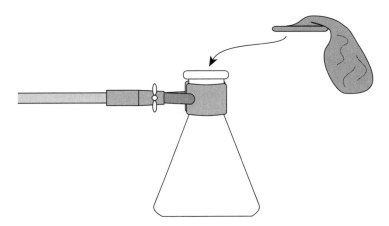

Figure 2: Have an assistant hold the flask with a laboratory clamp as you slip the balloon over the flask.

5. Allow the flask with the balloon attached to cool to room temperature.

 It may be helpful to guide the balloon as it is pushed into the flask to minimize the possibility of the balloon tearing. To speed up the cooling and therefore the pushing action, place the flask into ice water.

6. Have students record their observations.

7. (optional) Demonstrate that the balloon inflation and deflation can be repeated by returning the flask to the boiling water and warming it carefully as before. The flask can then be cooled.

 This procedure is not always successful as the balloon sometimes get stuck on the side of the flask and tears in the process. Be careful not to burn yourself with steam while heating the flask.

Part D: Boil Water in the Flask and Then Put the Balloon On

1. Prepare the balloon by stretching it several times.

2. (optional) Use a dropper to place about 10 drops of the soapy water solution in the balloon. The soapy solution acts as a lubricant.

 Take care not to get the soapy solution on the neck of the balloon; this can cause the balloon to slip off the neck of the flask.

3. Put enough water into the flask to cover the bottom (to a depth of about 3 mm, or ⅛ inch).

4. Heat the flask with the hot plate or burner and allow the water to boil until the flask is filled with steam. Do not let the flask boil dry.

5. Using tongs or gloves, remove the flask from the heat source.

Be careful when putting the balloon over the lip of the flask in the next step; steam can cause severe burns. It is helpful to use a ring stand and clamp or have an assistant carefully hold the flask with clamps, heat-resistant gloves, or tongs while the balloon is being stretched over the lip of the flask. Be sure the balloon is centered over the opening of the flask; if not, it will likely pop before it is completely drawn into the flask.

6. Carefully stretch the mouth of the balloon over the lip of the flask. (See Figure 1.)

7. Allow the flask to cool to room temperature.

➤ *The balloon will be pushed more dramatically and completely into the flask than in Part C. Again, it may be necessary to guide the balloon as it is being pushed into the flask to minimize the possibility of tearing. To speed up the cooling procedure and therefore the pushing action, place the flask in ice water.*

8. Have students record their observations.

9. (optional) Return the flask to the heat source and warm carefully. The balloon will be pushed out to its original position outside of the flask.

➤ *To be successful in this step, some water must remain inside the flask, and the balloon must not stick to the side of the flask. Sometimes the balloon gets stuck on the side of the flask and tears before it gets pushed out. The soap solution helps to prevent this.*

Part E: Develop an Explanation

1. Make a copy of the Observation Sheet into an overhead. Have students share their observations of the three balloon-flask systems prepared in Parts B–D. As students communicate their observations, fill in the figures on the overhead. Give students a chance to listen to and consider ideas proposed by other students.

➤ *Make sure that the figures you draw on the overhead closely match Figures A, C, and E on Overhead Masters 1–3.*

2. Use the top half of Overhead Masters 1–3 (provided) to provide a visual representation of the expected observations of the behavior of the balloons. Use the bottom half of each overhead to help the students to develop an explanation for these observations based on the behavior of the gaseous particles in the air. Discuss the figures on the Overhead Masters as a class. Be sure to discuss the following points:
   ○ Matter is conserved in a closed system.
   ○ Heating gaseous particles gives them more kinetic energy. This causes them to move faster and to hit the sides of the flask more frequently; this produces a greater force on a given area, so the larger pressure causes the balloon to inflate.
   ○ Cooling gaseous particles lessens their kinetic energy. This causes them to move more slowly and hit the sides of the flask less frequently. This produces a lesser force on a given area so the lesser pressure causes the balloon to collapse.
   ○ The expansion (or contraction) of the balloon results from an increase (or decrease) in the volume of empty space between the gaseous particles that make up the air. The size of the particles is NOT changed in the process.

Investigating Solids, Liquids, and Gases with **TOYS**

Have students relate this information to the evidence from their Observation Sheets.

EXTENSION

Use a chemical reaction to cause the balloon to be pushed into the flask. Soak steel wool in vinegar for several minutes. Use just enough vinegar to cover the steel wool. Remove the excess vinegar by squeezing the steel wool. Place the steel wool inside the flask. Stretch the mouth of the balloon over the opening of the flask as before. Let the flask sit overnight. The balloon will be pushed into the flask. (See the Explanation.)

• • • • • • • • • • • •

EXPLANATION

The following explanation is intended for the teacher's information. Modify the explanation for students as required.

Introducing the Activity challenges students to get a balloon into a flask. As might be expected, students typically suggest manually pushing the balloon into the flask, which works because nothing prevents the balloon from slipping into the flask. The challenge is then made more difficult when the balloon is first stretched over the mouth of the flask as shown in Figure 1. Students find that the air in the flask prevents the balloon from being pushed in—this is because air that is already in the flask takes up space. The Procedure then presents three different methods to try to get the balloon in the flask, each of which challenges the students to look at another aspect of the problem: Part B involves heating the flask with the balloon on it, Part C involves heating the flask before the balloon is placed on it, and Part D involves a phase change to get the balloon into the flask. To explain the observations that are made, let's look at the system on a molecular level.

Figures A and B on Overhead Master 1 provide a representation of the method presented in Part B. The dots in Figure B represent the gaseous particles that make up air. These gas particles are rapidly moving throughout the flask and the balloon (as denoted by the arrows in the figure). The pressure in the container is due to the particles hitting the walls of the container. Initially the pressure inside the flask is equal to that outside the flask (atmospheric pressure); however, as the flask warms, the pressure inside the flask increases.

Let's look at why the pressure increases inside the flask in Part B and what effect this pressure increase has on the behavior of the balloon. Heating the air in the flask causes the gaseous particles to gain kinetic energy. Thus, they move faster and collide with the walls of the container more frequently, thus increasing the pressure inside the flask to a pressure above atmospheric pressure. Because the balloon is stretchable, the increase in pressure inside the system causes the balloon to expand as the volume of the gas in the system increases. It is important to note that the number and size of the particles in the sealed flask remain constant. The volume increase results from an increase in the amount of empty space between the gaseous particles. Cooling the hot flask results in the balloon collapsing as the pressure and volume of the gas inside the system decrease and return to their original conditions.

In Part C, the gas in the flask is once again heated, but this time it's heated prior to the flask being sealed. (See Figures C and D on Overhead Master 2.) When an unsealed flask is heated, the gas inside the flask expands as before and some of the gas particles are pushed out of the flask. When the hot flask is then sealed with the balloon, the number of particles of gas inside the flask is fixed at a smaller number than would have been there prior to the heating procedure. When the sealed hot flask cools to room temperature, the particles move more slowly and strike the walls of the container less frequently, causing a reduction in pressure inside the flask. As this happens, the balloon is pushed into the flask because the atmospheric pressure is now greater than the pressure inside the flask. The volume of the gas inside the sealed flask is less than the original volume of gas since some of the gas was forced out during heating.

In Part D, the balloon is pushed into the flask once again as a result of the atmospheric pressure being greater than the pressure inside the flask, but this time a phase change creates the pressure difference and the observations are more dramatic. (See Figures E and F on Overhead Master 3.) Heating the liquid water causes it to turn into a gas (steam). The steam occupies a much larger volume in the flask than the liquid water. As a result, most of the air that was originally present is pushed out of the flask along with some of the steam. At this point, the flask is sealed and allowed to cool to room temperature. As the steam condenses back to liquid water, the amount of space it takes up is significantly less than the space the air occupies in Part C. Liquids occupy less than 0.1% of the space of a similar sample of a gas. (For example, steam at 100°C occupies a volume about 1,600 times the volume of the same mass of liquid water at 20°C.) There is very little gas above the liquid, and thus the gaseous pressure inside the sealed flask is much less than the atmospheric pressure. This pressure difference causes the balloon to be pushed into the flask.

A process similar to that in Part D occurs when fruits and vegetables are canned. As the jars cool, steam condenses and the gas remaining at the top of each jar contracts, which causes the pressure to decrease. Since the pressure of the atmosphere is greater than the pressure inside the jar, the higher atmospheric pressure on the lid seals the jar. The higher pressure outside the jar causes the metal lid to flex inward, causing a popping sound.

In the Extension, the steel wool soaked with vinegar reacts with oxygen to form rust (Fe_2O_3). This reaction consumes oxygen from the air. The consumption of oxygen inside the flask lowers the pressure inside the flask, causing the balloon to be pushed into the flask by atmospheric pressure. Because the rusting process is a slow one, the balloon is pushed into the flask more slowly than in Parts C and D of the Procedure. In addition, since air is only 20 percent oxygen, the change in pressure is much less than the change occurring when steam condenses to liquid water.

ASSESSMENT
Options:

- As an alternative to showing students Overhead Masters 1–3, have them create similar drawings as follows: For each part of the activity, have the students draw three to four sequential graphical representations of a model of what was occurring in the balloon-flask system, using dots for the particles of the gas and straight lines with arrowheads to indicate their movement. The number of particles represented should be 15 or less to avoid clutter. Pay particular attention to the numbers of particles represented in the sequence—matter (particles) should be conserved (except when an open flask is heated). Have the students then prepare a brief written explanation of the phenomena, referring to these graphics as appropriate. A sequence of plausible graphics is included on Overhead Masters 1–3.

- Evaluate students' completed Observation Sheets.

CROSS-CURRICULAR INTEGRATION
Home, safety, and career:
- Discuss why it is very dangerous to heat a closed glass container or incinerate aerosol cans.

REFERENCE

"Balloon Inside a Flask," *Fun With Chemistry: A Guidebook of K–12 Activities;* Sarquis, M., Sarquis, J., Eds.; Institute for Chemical Education: Madison, WI, 1991; Vol. 1, pp 155–158.

HANDOUT MASTERS

Masters for the following handouts are provided:
- Observation Sheet
- Overhead Master 1 for Part B
- Overhead Master 2 for Part C
- Overhead Master 3 for Part D

Copy as needed for classroom use.

Balloon-into-a-Flask Challenge
Observation Sheet

Record your observations in the table below.

Description of Balloon-Flask System	Draw your observations during the demonstration as indicated.		
	balloon on cool flask	balloon while heating	after cooling
balloon placed over unheated flask filled with air, flask heated and allowed to cool			
	heated flask (without balloon)	balloon on heated flask	after cooling
balloon placed over heated flask, allowed to cool			
	heated flask with boiling water (without balloon)	balloon on heated flask with water	after cooling
balloon placed over flask filled with water vapor, allowed to cool			

Balloon-into-a-Flask Challenge

Overhead Master 1 for Part B

Use the diagrams to explain the observations in this activity.

Figure A

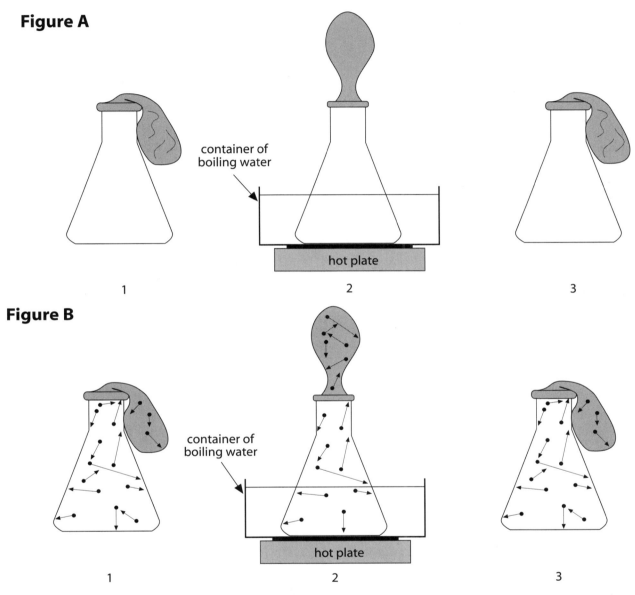

Figure B

container of
boiling water

hot plate

1

container of
boiling water

hot plate

2

3

At room temperature, the pressure inside the flask is equal to that outside the flask.

Heating the flask causes the air trapped inside it to expand. Expansion occurs because the gaseous particles making up the air have more kinetic energy and thus move faster when heated. Because the particles are moving faster, they collide with the walls of the container more often, which results in a higher pressure inside the container. This increased pressure causes the balloon to stretch, which allows the volume of the gas to expand. It is important to note that the number and size of the particles in a sealed flask remain constant.

When the flask cools to room temperature, the particles once again move slower and hit the walls of the container less often, resulting in a lower pressure inside the system. Therefore, the balloon collapses again.

Key
• = particles (molecules) of gas in the air

Balloon-into-a-Flask Challenge
Overhead Master 2 for Part C

Figure C

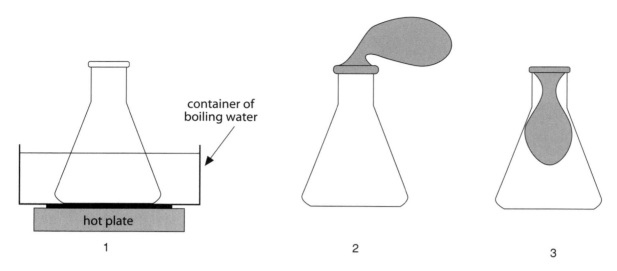

Figure D

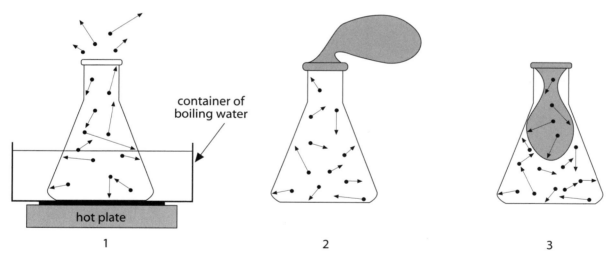

When an unsealed flask is heated, the gas inside expands as before, and some of the gas particles are pushed out of the flask.

When the hot flask is sealed with the balloon, the gas particles currently in the flask are trapped.

When the hot flask is allowed to cool to room temperature, the gas contracts (that is, there is a decrease in the volume of empty space between the gas particles). This causes a lower gas pressure inside the flask than out. The pressure difference causes the atmosphere to push the balloon down into the flask. Note that the size and number of gas particles inside the flask are the same as in Figure D.2.

Key
• = particles (molecules) of gas in the air

Balloon-into-a-Flask Challenge

Overhead Master 3 for Part D

Figure E

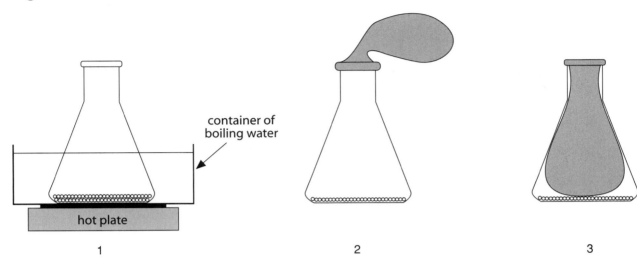

1

2

3

Figure F

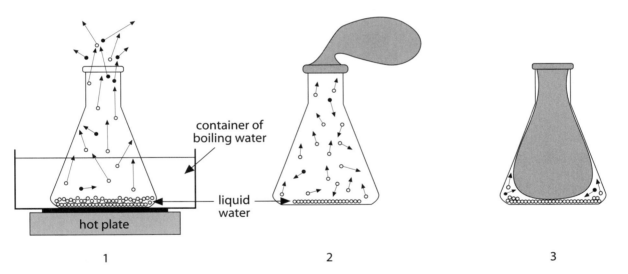

1

2

3

Boiling the liquid water causes some of it to turn into a gas (steam). The steam occupies a much larger volume in the flask than the liquid water. As a result, most of the air particles are pushed out of the flask along with some of the steam.

Seal the flask with the balloon. The makeup of the particle mixture in the flask is mostly water with just a little air.

After the flask is sealed with the balloon, it is allowed to cool to room temperature. Cooling allows the steam to condense back into a liquid; liquid water takes up much less of a volume than gaseous water. As a result, the pressure inside the flask is greatly reduced. The pressure of the atmosphere pushes the balloon down into the flask. Note the makeup of the particle mixture: Figure F.3 has the same amount of air as Figure F.2, much less steam than F.2, and more liquid water than F.2.

Key
  ○ = water molecules
  ● = particles (molecules) of gas in the air

Hands-On Activity

✔ **Demonstration**

Learning Center

Crushing an Aluminum Can

...Students see evidence of the volume decrease that occurs when a gas is condensed to a liquid.

The aluminum can is crushed.

✔ **Time Required**

Setup	15–20	minutes
Performance	20–25	minutes
Cleanup	5	minutes

✔ **Key Science Topics**

- change of state
- condensation
- pressure
- vaporization

✔ **Student Background**

Students should be familiar with the relationship between temperature and states of matter and know that a gas occupies a much larger volume than an equal mass of the same substance in the liquid or solid state.

✔ **National Science Education Standards**

Science as Inquiry Standards:

- Abilities Necessary to Do Scientific Inquiry

  Students differentiate description from explanation.

  Students review data from the experiment, summarize the data, and form a logical argument about the cause-and-effect relationships in the experiment.

  Students recognize and analyze alternative explanations proposed by other students.

Physical Science Standards:

- Properties and Changes of Properties in Matter

  A characteristic property of gases is that they expand when heated. A characteristic property of steam is that it occupies a much larger space than an equal mass of liquid water.

- Transfer of Energy

  Heat moves in predictable ways, flowing from warmer objects to cooler ones.

✔ **Additional Process Skills**

- predicting — Based on prior experience, students predict what will happen.

- defining operationally — Based on their observations, students state specific information about what causes the can to be crushed some times but not others.

MATERIALS

For Introducing the Activity
- (optional) dry ice, mittens or gloves, and goggles
- (optional) balloon

For the Procedure
Per class
- 4 aluminum soft-drink cans (clean and empty)
- hot plate

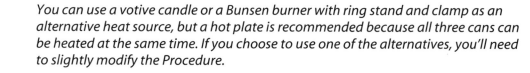

You can use a votive candle or a Bunsen burner with ring stand and clamp as an alternative heat source, but a hot plate is recommended because all three cans can be heated at the same time. If you choose to use one of the alternatives, you'll need to slightly modify the Procedure.

- tongs, beaker tongs, or heat-resistant gloves
- 1 of the following wide-mouthed containers:
  - large clear bowl with sides at least 5 inches high
  - wide-mouthed jar (approximately 1-quart)
  - beaker (400–600 mL)
  - pie or cake pan
  - "pop beaker" made from a cut-off plastic 2-L soft-drink bottle
- ice water
- goggles

For the Variation
Per class
- large, empty, clean can with small spout

A Collapsing Metal Can is available from Sargent-Welch® (#1513), P.O. Box 5229, Buffalo Grove, IL 60089-2362; 800/727-4368. Other cans must be thoroughly cleaned before use. See Safety and Disposal.

- rubber stopper or screw cap to fit the can spout
- hot plate
- hot pads
- (optional) container of water with crushed ice

SAFETY AND DISPOSAL

If using dry ice in Introducing the Activity, remember that dry ice is cold enough to cause frostbite or blisters. Use gloves and wear safety goggles whenever handling dry ice. Use in a well-ventilated area.

If doing the Variation, rinse the can thoroughly with water several times to ensure removal of any flammable substances originally in the can. The can and heat source will be hot; use care when handling. No special disposal procedures are required.

GETTING READY

1. Fill a wide-mouthed container ¾ full of ice water.

2. Fill one can with water.

3. Because the water in the full can will take much longer to boil than the water in the other can, place it on a heated hot plate 10–15 minutes before the activity is to be carried out. (Ideally a cloud of condensed water vapor should be observable above both of the water-containing cans in Part B, Step 2 of the Procedure.)

INTRODUCING THE ACTIVITY

If dry ice is available, show that dry ice occupies a much smaller volume than the same mass of gaseous carbon dioxide by placing a spoonful of crushed dry ice in a balloon. (See the activity "Using Dry Ice to Inflate a Balloon.") Be sure to protect your hands while filling the balloon, as dry ice can cause frostbite. Tie off the balloon and shake it to keep the dry ice from making the latex of the balloon brittle. Observe as the balloon expands due to the sublimation of the dry ice. Discuss the volume change that occurs. Alternatively, conduct the "Balloon-into-a-Flask Challenge" activity before doing this activity.

PROCEDURE

Part A: Organizing the Observations

Tell students that they will be observing attempts to crush four aluminum cans. Discuss the importance of making systematic observations during a scientific investigation so that the evidence collected can be used to propose explanations. Through discussion, clarify the difference between describing observations and proposing explanations. Show students the Data Sheet (provided) and explain that during the demonstrations, you want them to clearly describe the procedure and the outcomes they observe and to leave the "Explanation" column blank for now.

Part B: Crushing the Can

1. While students watch, pour about 5 mL water into the second can. Place this can and an empty can on the heated hot plate. (The can that is full of boiling or nearly boiling water should already be on the hot plate. See Getting Ready.) Tell the students that two of the cans contain water and one is empty (as a control), but do not alert the students to the difference in the volumes of water in the cans.

2. Heat the cans until the 5 mL water in the second can begins to boil. (This typically takes about 2–5 minutes to occur.) Condensed water vapor should be clearly visible above this can and the can that is full of water. Do NOT allow the can that contained 5 mL water to boil dry.

3. Ask the students to predict what will happen when the can that contained 5 mL water is lifted off the burner and inverted in cold water. Once predictions are made, use a pair of tongs to carefully lift the can from the hot plate. In one motion, invert the can and submerge its open end in the ice water in the wide-mouthed container.

➤ *If you hold the tongs in your hand with your palm pointed up, inverting the can will be easier. Be sure to keep the opening of the can below the ice-water level until it is crushed. A loud bang often accompanies the crushing of the can.*

4. Once the can is crushed, and with the students watching, lift the can above the water to show that a large amount of water drains out of the can. Ask the students where the water came from. (You may need to remind the students that this is the can that you poured about 5 mL water into initially.) *Much of the water that drained from the can was pushed into the can by the difference in pressure between the atmosphere and the inside of the inverted can. The pressure inside the can is reduced as the steam condenses to liquid water when the inverted can is cooled by the ice water.*

5. (optional) To reinforce the need for the water in the can to be heated to boiling, pour 5 mL water into an empty, unheated can. Repeat the inverting and cooling steps (Steps 3 and 4). (No change is observed, and the can is not crushed.) Ask students to explain the difference between the result with the can full of steam and the unheated can.

6. Repeat the inverting, cooling, and observing procedure (Steps 3 and 4) with the can that was empty. Challenge the students to explain the fact that the can is not crushed, but some water drains from the can when it is lifted. *The can is primarily filled with hot air. While the hot air is cooled, only a small volume change occurs, so the pressure difference between the inside of the can and the atmosphere isn't enough to crush the can. However, some water is pushed into the can due to the reduction in pressure associated with the decrease in temperature when the hot air is cooled by the ice water.*

7. Draw students' attention to the condensed water vapor coming from the can that is full of water. Ask them to predict what will happen when you invert this can in ice water. Without revealing the fact that the can is full of water, repeat the inverting, cooling, and observing procedure (Steps 3 and 4). Challenge the students to explain the observation. *The can is not crushed, because it is almost full of water. A large change in volume is not possible since the can is filled mostly with liquid water and has only a small volume of steam. The water that falls out of the can is the water that was originally in it.*

8. To help students form a clear understanding of the cause-and-effect relationship in these demonstrations, you may want to create a chart such as Table 1 in Assessment through a class discussion in which students contribute their observations.

Part C: Developing an Explanation

1. Challenge students to compare the four can-water systems from Part B. Have students propose explanations for the behavior of each system using evidence from their Data Sheet. Give students a chance to listen to and consider the explanations proposed by other students. They should remain open to and acknowledge different ideas and be able to accept the skepticism of others. Lead them in a discussion of the effect that a phase change has on the volume of the material.

2. Give students a chance to revise their explanations. Discuss as a class.

VARIATION

Larger Metal Can

1. Boil a small amount of water in the thoroughly rinsed large can. (See Figure 1.) Be sure all labels have been removed.

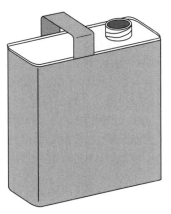

Figure 1: Boil a small amount of water in a thoroughly rinsed, large empty can.

2. Once steam has filled the can and is escaping rapidly, remove the can from the heat source.

3. Cap the can with a rubber stopper or a screw cap.

4. Allow the can to cool to room temperature or cool it in a container of water with crushed ice. The sides of the can will eventually be crushed inward as the can cools.

EXPLANATION

The following explanation is intended for the teacher's information. Modify the explanation for students as required.

When water is boiled, it changes from the liquid state to the gas state that we call steam. This process is called vaporization. As the steam forms, it drives much of the air out of the can. When the can is cooled, the steam changes back into liquid water. This process is called condensation. Vaporization and condensation are examples of a phase change—a change of a substance from one physical state to another.

When the opening of the can is sealed, either by inverting it under water or by capping it with a rubber stopper (see the Variation), air cannot enter the can. Water in the liquid state occupies considerably less volume than the same amount of water in the gaseous state, so when the steam condenses, the volume that it occupies is greatly decreased, and the pressure of gas inside the can decreases. The air outside the can exerts a pressure on the sides of the can that is greater than the pressure of the gas that remains inside the can. As a result, the can collapses. When the can is inverted in cold water, some of the cold water is also pushed into the can by the pressure of the atmosphere, but it cannot enter quickly enough to prevent the collapse of the can.

The can that contains only hot air is not crushed because the volume decrease from the cooling process is not sufficient to create the pressure difference required to collapse the metal can. The can that is filled with water is not crushed either, because cooling the liquid water in the can does not cause a significant decrease in the volume of the water. In addition, the volume of water vapor in the can is very small, so there is not a large change in volume as the water vapor condenses.

You can compare the force of the atmospheric pressure pushing the sides of the can inward to the force of a hammer denting the sides of a can. You may also wish to allow students to see if they can duplicate the effects by simply squeezing the aluminum can with tongs. Do not allow students to squeeze the cans with their bare hands; sharp edges may cause severe cuts.

ASSESSMENT

• Have students turn in their completed Data Sheets. Sample data is included in Table 1.

Table 1: Sample Data for Part B			
Contents of Can Before Heating	**Contents of Can After Heating**	**Result when Can is Inverted in Cold Water**	**Explanation**
5 mL water and air	steam, some water	The can collapses with a bang, and a fair amount of water is pushed into the can.	Condensing steam creates a partial vacuum, atmospheric pressure causes the can to collapse and water to be pushed into the can.
5 mL water and air	unheated; contains water, air	The can does not collapse.	No steam is present to condense, so the pressure inside the can remains the same as the atmospheric pressure. A little extra water may be pushed into the can by the force of the inversion.
air only	hot air	The can does not collapse, and some water is pushed into the can.	Results are due to the volume decrease as the hot air is cooled.
full of water	hot water	No apparent change.	The can is already full of water, so no apparent volume change in the water is noted.

REFERENCES

Bell, J., "Spectacular Chemical Demonstrations—Without Flames, Explosions, or Mess," demonstration at the American Chemical Society Division of Chemical Education Biennial Conference, Bucknell University, Lewisburg, PA, 1994.

"Collapsing Aluminum Cans"; *Fun with Chemistry: A Guidebook of K–12 Activities*; Sarquis, M., Sarquis, J., Eds.; Institute for Chemical Education: Madison, WI, 1993; Vol. 1, pp 159–162.

CONTRIBUTORS

Larry Berg, Eastern Green County Elementary School, Bloomfield, IN; Teaching Science with TOYS, 1995.

David Burch, Eastern Green County Junior-Senior High School, Bloomfield, IN; Teaching Science with TOYS, 1995.

Rosa Hernandez, Rosemont Avenue School, Los Angeles, CA; Teaching Science with TOYS, 1996.

Chris Mihm, Rosemont Avenue School, Los Angeles, CA; Teaching Science with TOYS, 1996.

HANDOUT MASTER

A master for the following handout is provided:
- Data Sheet

Copy as needed for classroom use.

Crushing an Aluminum Can
Data Sheet

Record your observations in the table below.

Contents of Can Before Heating	Contents of Can After Heating	Result when Can is Inverted in Cold Water	Explanation
5 mL water and air			
5 mL water and air	unheated		
air only			
full of water			

Hats Off to the Drinking Bird

...What makes the drinking bird so thirsty? Will it ever stop drinking? Observe the behavior of the drinking bird and use scientific inquiry to understand this toy.

✔ Time Required

Setup   5 minutes
Performance   20–30 minutes*
Cleanup   5 minutes
*Additional time is required if done as a hands-on activity.

✔ Key Science Topics

- center of gravity
- effect of temperature on gas pressure
- energy and phase changes
- evaporative cooling and rate of evaporation
- liquid-vapor equilibrium

Drinking bird

✔ Student Background

This activity is most effective after students have studied the states of matter and have some understanding of changes of state. Specifically, students should understand that evaporation has a cooling effect because high energy molecules become vapor, leaving the remaining liquid molecules with a lower average energy. An understanding of the center of gravity is recommended.

✔ National Science Education Standards

Science as Inquiry Standards:

- Abilities Necessary to Do Scientific Inquiry

  Students identify questions that can be answered through scientific investigations.

  Students develop explanations using evidence obtained from their investigations.

  Students listen to and acknowledge the explanations proposed by other students.

Physical Science Standards:

- Properties and Changes of Properties in Matter

  The characteristic properties of the liquid and gas contained in the bird lead to its distinctive behavior.

- Transfer of Energy

  Heat is transferred in predictable ways. The evaporation of liquid from the head of the bird cools the gas inside.

✔ Additional Process Skill

- predicting   Students predict the effect that changing temperature and humidity will have on the behavior of the drinking bird.

257

MATERIALS

For the Procedure
Per demonstration or group
- drinking bird
- cup of room-temperature tap water high enough for the bird's beak to dip into the water
- cotton balls or cotton-tipped swabs
- (optional) drinking bird with felt removed from the head (See Getting Ready.)

For Variations and Extensions

❷ All materials listed for the Procedure plus the following:
Per demonstration or group
- cup of rubbing alcohol (70% isopropyl alcohol solution)
- cup of hot tap water
- cup of cold water (or ice cubes and tap water)
- masking tape and marker to label cups

❹ All materials listed for the Procedure plus the following:
Per demonstration or group
- 1 of the following:
  - large jar
  - zipper-type plastic bag
  - covered aquarium
- cup of rubbing alcohol

❻❼ All materials listed for the Procedure plus the following:
Per demonstration or group
- stopwatch
- graph paper

❽ All materials listed for the Procedure plus the following:
Per demonstration or group
- weights such as coins
- tape

SAFETY AND DISPOSAL

The drinking bird is made of thin glass and is fragile. Care should be taken not to break it because it contains freon or methylene chloride. Freon should not be released into the atmosphere because of its environmental effect on the ozone in the upper atmosphere. Methylene chloride is toxic; if a drinking bird is broken, avoid breathing fumes or allowing contact with skin. Rubbing alcohol is intended for external use only. No special disposal procedures are required.

GETTING READY

It is imperative that the drinking bird be tested before being used in class.

Often some adjustment is needed for the bird to bob properly. The pivot point must be in the correct position for the bird to bob, and the tube has a tendency to slide out of adjustment, which prevents proper action. Sometimes the metal supports need to be bent so that the bird's head sits slightly forward of the support points. Also, the height of the water container must be adjusted so that the liquid in the top chamber flows back into the bottom chamber when the bird bobs. On a humid day, it will be harder to get the bird to bob, so these adjustments become even more critical.

Label cups of room-temperature water, hot water, cold water, and rubbing alcohol for Extension 2. If done as a hands-on activity, different student groups can be assigned different temperatures of water to investigate.

If desired, carefully remove the exterior felt (without removing the beak) from the head of a drinking bird.

INTRODUCING THE ACTIVITY

Show students a drinking bird in action without providing an explanation. You may wish to display the bird for one or more days to arouse curiosity. Explain to the students that they will be using scientific inquiry to discover how the drinking bird works.

PROCEDURE

1. Tell students that they will be carefully observing the drinking bird and writing down their observations. Explain that this is the first step in their scientific inquiry—they must make systematic observations that will help them pose meaningful hypotheses later. Emphasize that they are not to make hypotheses or inferences at this time—only to write down observations.

2. Allow students time to write down their observations and descriptions of the drinking bird phenomenon on the Instruction and Observation Sheet (provided).

3. Discuss student observations at this point; again emphasize the difference between observations and inferences. Before proceeding further, make sure that all students have seen the complete cycle of the bird and have noted the flow of the colored liquid in the drinking bird.

4. Ask the class for hypotheses as to why the bird bobs. Emphasize that their hypotheses should be logical outcomes of their observations and must lead to questions that can be answered through scientific investigation.

5. Show students a drinking bird with a dry head or no felt on its head. Manually pivot the bird forward about as far as the bird with the water cup dips to drink. Have students observe carefully and discuss. Does the behavior of this bird support their hypotheses? How? If not, how would they change their hypotheses? Make sure students listen to and consider the explanations proposed by their classmates and also remain open to the skepticism of others.

6. Ask, "What else can we do to test your ideas?" Experiment with student suggestions or ask them guided questions such as:

   a. "Will the bird bob if the head is soaked in hot water?" *Yes.*

   b. "If the tail is warmed, does the liquid rise faster?" *Yes.* (Apply heat by cupping your hand around the lower bulb or placing the lower bulb in warm water.)

 Do not heat with a flame as the thin glass may break.

   c. "If the tail is cooled (by dipping in cold water), what happens to the level of the liquid in the tube?" *It is lowered.*

7. Tell students that you are going to provide another clue to the behavior of the drinking bird. Use a cotton ball or swab to place a small amount of water on the back of the students' hands. This will remind them that as a liquid evaporates, cooling occurs.

8. Facilitate the discussion, helping students to conclude that a temperature difference between the head and tail sections is the driving force of the bird's action. (This temperature difference is primarily a result of cooling by evaporation. The rate of bobbing is faster when warm water is used, since the rate of evaporation is faster.)

• • • • • • • • • • • • VARIATIONS AND EXTENSIONS

1. Set up the activity in a learning center and have students conduct the investigation using the Instruction and Observation Sheet (provided).

2. Have different groups test their drinking birds using different temperatures of water or using rubbing alcohol.

 If alcohol is used, the beak has a tendency to come unglued and fall off.

3. Have students determine how long the bird will continue to bob once the cup of water is removed.

4. Challenge students to devise ways of increasing or decreasing the rate of bobbing. Lead them to try the following:

   a. To increase the rate of evaporation, fan or gently blow air across the head (CAUTION—the bird is fragile) or use a more volatile liquid, such as rubbing alcohol, in place of the water.

   b. To decrease the rate of bobbing, prevent evaporation by placing the bird in a large jar, a large zipper-type plastic bag, or a covered aquarium; after a while, the bird stops bobbing because once the atmosphere is saturated with water vapor, evaporation (and thus cooling) stops.

5. Challenge students to design a bobbing bird without the use of any exterior liquid. (See the Explanation and the Plumb, 1975 reference.)

6. Have students count the number of bird bobs per minute under various conditions and graph the results.

7. Have students count the number of bird bobs per minute on a day with high humidity and on a day with low humidity. Is there a difference?

 If humidity is very high, the bird may bob more slowly or not at all unless fanned.

8. Show the effects of lowering the center of gravity on the stability of the bird. Taping weights such as coins at or near the bottom of the bird makes the bird remain stable instead of tipping as the liquid rises in the neck.

 ## EXPLANATION

The following explanation is intended for the teacher's information. Modify the explanation for students as required.

The drinking bird is a toy that, once set into motion, dips its beak into a glass of water (or alcohol), returns to its upright position, then dips its beak into the liquid again. This bobbing pattern is repeated for an extended period of time. Initially, the bird is in an upright position, shown in Figure 1.

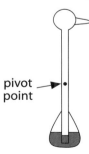

pivot point

Figure 1: Initially, the drinking bird is in the upright position.

Note the level of liquid in the bottom chamber and in the tube that is the neck of the bird. Initially, the levels are about the same, indicating that the pressure in the upper chamber (the head) and the bottom chamber (the tail) are about the same. When the felt head is placed into the glass of water, the water is absorbed by the felt beak and head as a result of capillary action. As the water evaporates, the upper chamber is cooled. This cooling lowers the pressure of the vapor in the upper chamber.

Since the lower chamber remains at room temperature, the pressure of vapor in the lower chamber remains the same. Because the pressure in the upper chamber has decreased, the colored liquid in the lower chamber is pushed up into the neck (moving from higher pressure to lower pressure) as illustrated in Figure 2.

Figure 2: The colored liquid in the lower chamber is pushed up into the neck.

As more liquid is pushed from the bottom chamber, it begins to fill the upper chamber. This raises the center of gravity above the pivot point, and the bird becomes unstable and starts to tip toward its "drinking" position. (See Figure 3.)

Figure 3: The bird becomes unbalanced and starts to tip toward its drinking position.

When the bird tips forward until it is almost horizontal, it once again gets its beak wet ("drinks"), replacing the water that has evaporated. At this point the tube opening in the tail is above the surface of the liquid in the tail. This allows gas from the tail to rise into the neck towards the head and liquid to run back into the tail. (See Figure 4.) The redistribution of the liquid causes a shift in the center of gravity, and the bird returns to the upright position, essentially the same configuration as in Figure 1. More evaporation takes place, and the cycle is repeated.

Figure 4: The liquid in the drinking bird is redistributed.

On humid days (see Extension 7) or in an isolated atmosphere such as a covered aquarium (see Extension 4), the action stops as the air becomes saturated with water vapor and evaporation stops. When the bird drinks alcohol, the rate of bobbing generally is faster. This happens because alcohol is more volatile, evaporates at a faster rate, and thus cools the vapor in the head more rapidly.

Without the felt on the bird's upper chamber, water is not absorbed by the glass bulb, and evaporation does not take place; without evaporation and the associated cooling, the bird does not bob.

A "waterless" bobbing bird can be made by painting the bottom chamber black and the upper chamber white. With this design, the difference in absorption of radiant heat from the sun or a light bulb causes the temperature difference and therefore the pressure difference instead of evaporation.

The bird can be viewed as a heat engine, a device that utilizes a working fluid to exchange heat with two heat reservoirs and perform work. The bird's liquid is the working fluid. It absorbs heat in the lower chamber at room temperature and gives off heat during evaporation at the cooler head. There is a net absorption of heat; therefore, the bird has the capacity to do work, initially in the form of gravitational potential energy of the liquid. In normal operation, this potential energy is converted into the kinetic energy of the observed bobbing motion, although some mechanism could be designed to have the bird do work as this motion takes place.

CROSS-CURRICULAR INTEGRATION

Life science:
• Discuss how the human body cools itself through evaporation (sweating).

REFERENCES

Frank, D. "The Drinking Bird and the Scientific Method," *Journal of Chemical Education.* 1973, *50,* 211.

Juergens, F. University of Wisconsin-Madison, personal communication.

Plumb, R. "Physical Chemistry of the Drinking Duck," *Journal of Chemical Education.* 1973, *50,* 212.

Plumb, R. "Footnote to the Drinking Duck Exemplum," *Journal of Chemical Education.* 1975, *52,* 728.

CONTRIBUTORS

Alison Dowd, Talawanda Middle School, Oxford, OH; Teaching Science with TOYS peer writer.

Gary Lovely, Edgewood Middle School, Hamilton, OH; Teaching Science with TOYS peer mentor.

Tom Runyan, Garfield Alternative School, Middletown, OH; Teaching Science with TOYS peer mentor.

HANDOUT MASTER

A master for the following handout is provided:

- Instruction and Observation Sheet

Copy as needed for classroom use.

Name _____ Date _____

Hats Off to the Drinking Bird
Instruction and Observation Sheet

1. Carefully observe the drinking bird in action and write down your observations.

    _____

    _____

    _____

    _____

2. Based on your observations, construct a hypothesis as to why the bird bobs.

    _____

    _____

    _____

    _____

3. Now, carefully observe a drinking bird with a dry head or no felt on its head. Does the behavior of this bird support your hypothesis? How? If not, how would you change your hypothesis?

    _____

    _____

    _____

    _____

4. How can you test your ideas? _____

    _____

    _____

    _____

5. Based on your testing, discuss why the drinking bird bobs. (Use the back of this sheet if needed.)

    _____

    _____

    _____

    _____

Appendix A:

Masters for Assessment Models

This Appendix contains masters for the following materials. See Pedagogical Strategies for a discussion of how to integrate these materials with the lessons.

- Performance Evaluation
- Pre- and Post-Test
- Five-Minute Videotape Production Performance Evaluation
- Concept Poster Performance Evaluation
- Concept Cartoon Performance Evaluation
- Concept Song Performance Evaluation

Performance Evaluation

Activity _____

Name(s) _____

Period _____ **Date** _____

Performance Rating:	No Evidence	0 pts
	Approaches Goal	1 pt
	Meets Goal	2 pts
	Exceeds Goal	3 pts
	Not Applicable	N/A

Demonstrates independence in the interpretation and the performance of the procedures _Remarks:_	0	1	2	3	N/A
Demonstrates appropriate use of equipment _Remarks:_	0	1	2	3	N/A
Demonstrates the ability to collect data _Remarks:_	0	1	2	3	N/A
Demonstrates the ability to make accurate measurements and to judge the reasonableness of the results _Remarks:_	0	1	2	3	N/A
Demonstrates an understanding of the concepts involved in the activity _Remarks:_	0	1	2	3	N/A
Properly uses mathematics, formulas, conversions, and units to solve the problem _Remarks:_	0	1	2	3	N/A
Analyzes the results and forms appropriate conclusions to answer teacher follow-up questions _Remarks:_	0	1	2	3	N/A
Demonstrates safe laboratory practices _Remarks:_	0	1	2	3	N/A
Works effectively as a team _Remarks:_	0	1	2	3	N/A

Pre- and Post-Test

Activity _____

Name(s) _____

Period _____   **Date** _____

1. Identify the state of matter illustrated in each picture and explain your answers.

_____   _____   _____

_____   _____   _____

_____   _____   _____

_____   _____   _____

_____   _____   _____

2. Complete the table below. In the first column, use a different state of matter for each line. When describing the volume and shape, use the terms _definite_ or _not definite_.

State	Volume	Shape	Movement of Molecules
			random motion; not touching
	definite	definite	
		not definite	

3. You take a Popsicle™ out of the freezer and lay it on the kitchen table. When you return you notice the wrapper is not firm. When you open it, liquid runs out.

   a) What was the original state of the Popsicle? _____

   b) What was the final state of the Popsicle?_____

   c) What change of state occurred? _____

4. It is a hot and sunny day. You get very warm and your sweat glands start to produce perspiration. How does the perspiration help you maintain a stable body temperature?

   _____

   _____

   _____

   _____

   _____

5. You buy a cold soft drink from a machine. The can is dry when it first comes out of the machine, but in a few minutes you notice that droplets of water form on the outside of the cold can.

   a) Where did the water come from?

   b) What change of state is occurring?

   c) What state was the water in before it formed the droplets of water?

Reproduced from *Investigating Solids, Liquids, and Gases with TOYS*, published by Terrific Science Press.

Five-Minute Videotape Production Performance Evaluation

Activity _____

Name(s) _____

Period _____   **Date** _____

Performance Rating:	No Evidence	0 pts
	Approaches Goal	1 pt
	Meets Goal	2 pts
	Exceeds Goal	3 pts
	Not Applicable	N/A

Demonstrates an understanding of the concept illustrated in the video *Remarks:*	0 1 2 3 N/A			
Selects appropriate scenes to illustrate the concept *Remarks:*	0 1 2 3 N/A			
Uses creativity in selecting the scenes *Remarks:*	0 1 2 3 N/A			
Video is well produced *Remarks:*	0 1 2 3 N/A			

Concept Poster Performance Evaluation

Activity _____

Name(s) _____

Period _____ **Date** _____

Performance Rating:	No Evidence	0 pts
	Approaches Goal	1 pt
	Meets Goal	2 pts
	Exceeds Goal	3 pts
	Not Applicable	N/A

Demonstrates an understanding of the
concept illustrated in the poster
Remarks: 0   1   2   3   N/A

Selects appropriate pictures/drawings to
illustrate the concept
Remarks: 0   1   2   3   N/A

Uses creativity in drawing the pictures or
selecting the pictures
Remarks: 0   1   2   3   N/A

Poster is neat and includes appropriate titles
Remarks: 0   1   2   3   N/A

Demonstrates an understanding of the concept
during the in-class presentation
Remarks: 0   1   2   3   N/A

Presents in a clear and accurate manner
Remarks: 0   1   2   3   N/A

Concept Cartoon Performance Evaluation

Activity _____

Name(s) _____

Period _____   **Date** _____

Performance Rating:	No Evidence	0 pts
	Approaches Goal	1 pt
	Meets Goal	2 pts
	Exceeds Goal	3 pts
	Not Applicable	N/A

Demonstrates an understanding of the concept illustrated in the cartoon
Remarks: 0   1   2   3   N/A

Selects appropriate pictures/drawings to illustrate the concept
Remarks: 0   1   2   3   N/A

Selects appropriate story line to illustrate the concept
Remarks: 0   1   2   3   N/A

Uses creativity in drawing the pictures
Remarks: 0   1   2   3   N/A

Uses creativity in the story line
Remarks: 0   1   2   3   N/A

Demonstrates an understanding of the concept during the in-class presentation
Remarks: 0   1   2   3   N/A

Presents in a clear and accurate manner
Remarks: 0   1   2   3   N/A

Concept Song Performance Evaluation

Activity _____

Name(s) _____

Period _____ **Date** _____

Performance Rating:	No Evidence	0 pts
	Approaches Goal	1 pt
	Meets Goal	2 pts
	Exceeds Goal	3 pts
	Not Applicable	N/A

Demonstrates an understanding of the concept illustrated in the song
Remarks:

0   1   2   3   N/A

Selects appropriate words to represent the concept
Remarks:

0   1   2   3   N/A

Uses creativity in writing the song
Remarks:

0   1   2   3   N/A

Song is clear, understandable, appropriate
Remarks:

0   1   2   3   N/A

Shopping List

The activities in this book primarily use common materials available at most discount, grocery, or hardware stores. However, several lessons use materials (some of which are optional) that must be ordered from special suppliers. We provide ordering information in each lesson; the list below is for your convenience in preparing a "shopping list" of items you will probably need to order in advance. This is not an exhaustive list of all materials required; complete lists are provided in the activities.

Suppliers

Fisher Scientific	http://www.fishersci.com	800/955-1177
Flinn Scientific	http://www.flinnsci.com	800/452-1261
Sargent-Welch	http://www.sargentwelch.com	800/727-4368
Toysmith	http://www.toysmith.com	800/356-0474

Lessons Requiring Ordered Items

Properties of Matter
- syringe needle, #14-826-5B; Fisher Scientific
- syringe, #14-823-2D; Fisher Scientific

Marshmallow in a Syringe
- syringe needle, #14-826-5B; Fisher Scientific
- syringe, #14-823-2D; Fisher Scientific

Non-Newtonian Fluids—Liquids or Solids?
- polyvinyl alcohol (powder form), #P0153 for 100 g, **or**
  polyvinyl alcohol (4% aqueous solution), #P0209 for 500 mL; Flinn Scientific

Crystals from Solutions
- Magic Tree kit, #8309; Toysmith
- 100-mm x 15-mm Petri dish, #08-757-12; Fisher Scientific

Boiling Liquids in a Syringe
- syringe needle, #14-826-5B; Fisher Scientific
- syringe, #14-823-2D; Fisher Scientific
- (optional) Luer Tip cap, #14-826-76; Fisher Scientific

Crushing an Aluminum Can
- Collapsing Metal Can #1513; Sargent-Welch (for a Variation)

The Phase Changes of Carbon Dioxide
- wide-stem Beral pipet, #AP2253 for 20 or #AP8480 for 500; Flinn Scientific

National Science Education Standards Matrix

This matrix shows how the activities in this book relate to the National Science Education Standards. The standards are taken from *National Science Education Standards;* National Research Council; National Academy: Washington, D.C., 1996.

	Activities			
	Balloon in a Bottle	Balloon-into-a-Flask Challenge	BedBugs	Boiling Liquids in a Syringe
Science as Inquiry Standards				
Abilities Necessary to Do Scientific Inquiry				
Identify questions that can be answered through scientific investigations.				✔
Design and conduct a scientific investigation.	✔			✔
Use appropriate tools and techniques to gather, analyze, and interpret data.				
Develop descriptions, explanations, predictions, and models using evidence.		✔	✔	
Think critically and logically to make the relationships between evidence and explanations.		✔	✔	
Recognize and analyze alternative explanations and predictions.				
Communicate scientific procedures and explanations.		✔		
Physical Science Standards				
Properties and Changes of Properties in Matter				
A substance has characteristic properties, such as density, a boiling point, and solubility, all of which are independent of the amount of the sample. A mixture of substances often can be separated into the original substances using one or more of the characteristic properties.	✔	✔	✔	✔
Substances react chemically in characteristic ways with other substances to form new substances (compounds) with different characteristic properties. In chemical reactions, the total mass is conserved. Substances often are placed in categories or groups if they react in similar ways; metals is an example of such a group.				
Chemical elements do not break down during normal laboratory reactions involving such treatments as heating, exposure to electric current, or reaction with acids. There are more than 100 known elements that combine in a multitude of ways to produce compounds, which account for the living and nonliving substances we encounter.				
Motions and Forces				
The motion of an object can be described by its position, direction of motion, and speed. That motion can be measured and represented on a graph.		✔		
Light, Heat, Electricity, and Magnetism				
Energy is a property of many substances and is associated with heat, light, electricity, mechanical motion, sound, nuclei, and the nature of a chemical. Energy is transferred in many ways.			✔	
Heat moves in predictable ways, flowing from warmer objects to cooler ones, until both reach the same temperature.		✔		

Activities

Boiling Water in a Paper Pot	Boiling Water with Ice	Burping Bottle	A Cool Phase Change	Crushing an Aluminum Can	Crystals by Freezing	Crystals from Solutions	Disappearing Air Freshener	Hats Off to the Drinking Bird	Liquid to Gas in a Flick	Marshmallow in a Syringe	Moving Molecules	Mystery Eggs	Non-Newtonian Fluids—Liquids or Solids?	The Phase Changes of Carbon Dioxide	Properties of Matter	Rock Candy Crystals	Showing That Air Has Mass	Tissue in a Cup	Using Dry Ice to Inflate a Balloon
		✔						✔											
		✔	✔		✔	✔	✔						✔	✔	✔	✔			
		✔	✔						✔				✔		✔				
✔				✔	✔	✔	✔	✔		✔	✔				✔			✔	
	✔			✔						✔					✔		✔		✔
				✔				✔											
													✔						
✔	✔	✔	✔		✔	✔	✔	✔	✔	✔	✔	✔		✔	✔	✔	✔	✔	✔
													✔						
												✔							
✔			✔	✔	✔			✔											